PLATELET PROTOCOLS

PLATELET PROTOCOLS

RESEARCH AND CLINICAL
LABORATORY PROCEDURES

Melanie McCabe White, B.A.

and

Lisa K. Jennings, Ph.D.

Department of Medicine
University of Tennessee, Memphis

Selected Illustrations by Michael P. Condry

Academic Press
San Diego   New York   Boston
London   Sydney   Tokyo   Toronto

Academic Press
a division of Harcourt Brace & Company
525 B Street, Suite 1900, San Diego, California 92101-4495
http://www.academicpress.com

Academic Press Limited
24-28 Oval Road, London NW1 7DX, UK
http://www.hbuk.co.uk/ap/

Library of Congress Catalog Card Number: 98-83126

Printed and bound in the United Kingdom
Transferred to Digital Printing, 2011

CONTENTS

Contents

PREFACE

Platelets are required for hemostasis and are key participants in pathologic thrombosis. Individual platelets circulate in an unactivated state. In response to vascular injury, platelets adhere to the subendothelium, where adhesion and platelet activation lead to formation of the thrombus. When hemostasis is initiated by a pathologic event such as plaque rupture, thrombus formation can lead to vasoclusion, ischemia, or infarction. A growing body of knowledge in hemostasis and thrombosis over the last twenty years has demanded a current handbook on the evaluation of platelet function in both the basic research and clinical laboratories.

In this book we have endeavored to provide a systematic review of key points important in evaluation of platelet function and to outline the methods necessary for proper platelet function testing. The first chapter is a basic introduction to platelets. The major platelet membrane proteins, the events associated with platelet aggregation, and the fundamental platelet agonists are described. The second chapter outlines laboratory evaluation of platelet function, including the variables that can affect the platelet aggregation response, the proper methods for collecting and processing blood for platelet aggregation, methods for measuring platelet aggregation and secretion, as well as other platelet function testing. Chapter three is an introduction to the functional abnormalities of platelets. This is not intended to serve as a comprehensive fund of knowledge, but it does provide key information with regard to abnormalities that can affect the platelet response. The final chapter introduces anti-platelet drugs that have an effect on platelet function, including the newly developed ticlopidine and clopidogrel, as well as the GPIIb–IIIa antagonists. A discussion of the pharmacodynamic effects of these drugs is included along with the criteria for evaluating their effect on platelet function.

This book will provide information for evaluation of platelet function that is timely and logically organized. The text will be useful to basic science investigators, laboratory staff, and clinicians who evaluate or manage patients in the area of thrombosis and hemostasis. We also hope to satisfy the great new demand for a text that concisely outlines the current protocols for platelet function evaluation which has arisen with the rapid development of the new anti-platelet therapies.

Lisa Kyle Jennings
Melanie McCabe White

Acknowledgments

A sincere thank you to Dr. David R. Phillips for giving me the opportunity to learn and value scientific inquiry.

L.K.J.

I would like to thank Dr. Mervyn A. Sahud for introducing me to the wonders of platelets in 1972 and for his continued mentorship through the ensuing thirteen years.

M.M.W.

CHAPTER

1

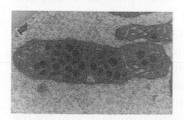

BASIC
INTRODUCTION
TO PLATELETS

INTRODUCTION

Platelets are anucleate cytoplasmic fragments derived from human bone marrow megakaryocytes. They have a seminal role in hemostasis and thrombosis. Human platelets in circulation are discoid in morphology and have a relatively smooth surface. The ultrastructure of the normal platelet reveals the surface-connected canalicular system (sccs), which is a continuous membrane system extending from the plasma membrane (pm) and the dense tubular system (dts) (Figure 1.1). The plasma membrane contains many proteins that serve as receptors for agonists that initiate platelet activation responses or for adhesive molecules that mediate platelet adhesion and platelet aggregation. The surface-connected canalicular system pro-

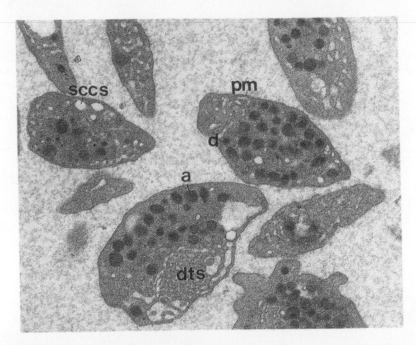

Figure 1.1. Transmission electron micrograph of normal resting platelets (×15,000). pm, plasma membrane; dts, dense tubular system; sccs, surface-connected canalicular system; d, dense granules; a, alpha-granules.

vides a passage for plasma components as well as for vasoactive substances that are released upon platelet activation. Platelets contain several types of storage granules (d and a), the contents of which are released upon platelet activation. The dense tubular system has been shown to be the site from which the calcium is released that triggers many of the calcium-dependent activation enzymes and contractile events in the platelet activation response.

PLATELETS IN HEMOSTASIS

Damage to the vascular endothelium usually leads to exposure of the subendothelial layer, which encourages adherence of platelets through various receptors but primarily through GP (glycoprotein) Ib–IX–V. Adherent platelets become activated and release the contents of storage granules, recruiting nearby platelets in circulation to form an aggregate (Figure 1.2). The formation of the platelet aggregate or thrombus occurs via activation of GPIIb–IIIa and binding of multivalent adhesive ligands, fibrinogen, or von Willebrand factor (vWF), which crosslink the adjacent activated platelets (Figure 1.3). An aggregate, as shown in Figure 1.3B, is an amorphous mass of many platelets. The stronger the agonist, the larger the aggregate size. When platelets are unactivated, the GPIIb–IIIa is in a low affinity state and cannot bind soluble fibrinogen. Ligand binding to activated GPIIb–IIIa is regulated by platelet activation. Upon platelet activation, GPIIb–IIIa undergoes a conformational change that permits fibrinogen to bind. Fibrinogen contains at least three platelet recognition sites: the Arg–Gly–Asp–Phe (RGDF) and the Arg–Gly–Asp–Ser (RGDS) at Aα, and the dodecapeptide sequence His–His–Leu–Gly–Gly–Ala–Lys–Gln–Ala–Gly–Asp–Val (HHLGGAKQAGDV) at the γ400–411 region of fibrinogen. Even though the fibrinogen molecule has two RGDX sequences, fibrinogen binding to the platelet appears to be primarily mediated through the dodecapeptide sequence. Due to their small size, synthetic peptides of these three sequences bind GPIIb–IIIa regardless of the GPIIb–IIIa

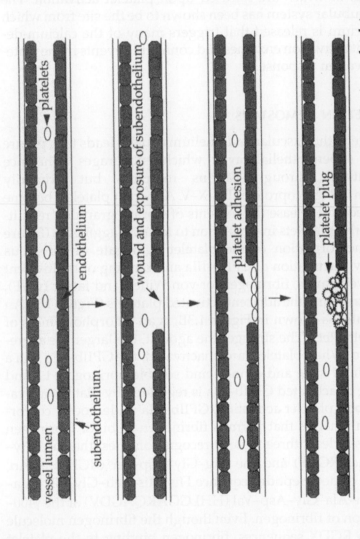

Figure 1.2. Platelet plug formation in response to vascular injury. Upon vessel injury, platelets adhere to the exposed subendothelium. Platelet activation and platelet aggregation occur leading to thrombus formation.

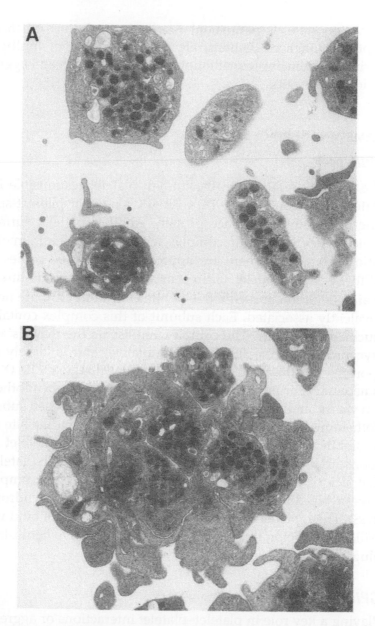

Figure 1.3. Transmission electron micrograph of resting discoid platelets (A) and platelet aggregates (B). Note the presence of numerous dense granules in the resting discoid platelets.

activation state. Experimental results from the use of these peptide sequences demonstrate the importance of multiple fibrinogen domains in mediating both the adhesion and aggregation of platelets.

Membrane Surface Proteins

GPIb–IX–V

The primary receptor on the human platelet responsible for platelet adhesion is GPIb–IX. GPIb–IX is a major platelet-specific heterodimeric surface protein. Another platelet surface protein, GPV, is found in complex with GPIb–IX at a 1:2 molar ratio (GPV:GPIb). There are approximately 25,000 copies of GPIb–IX–V per platelet. GPIb consists of two disulfide-linked chains — GPIbα and GPIbβ — while GPIX and GPV are noncovalently associated. Each subunit of this complex contains leucine-rich repeats. The complex represents the major sialoglycoprotein on the surface of the platelet and is largely responsible for the platelet surface negative charge. GPIb–IX–V is necessary for normal platelet adhesion to the subendothelium via its binding to immobilized vWF on the exposed subendothelium. In addition, GPIb–IX–V contains a binding site for thrombin and has been implicated in enhancing platelet response to low thrombin concentrations. Patients with platelets that lack GPIb–IX–V or have poor expression of this complex have a bleeding disorder called Bernard–Soulier syndrome (BSS). These platelets are unusually large and do not bind von Willebrand factor, thus failing to form an effective hemostatic plug.

GPIIb–IIIa

Playing a key role in platelet–platelet interactions or aggregation is glycoprotein GPIIb–IIIa, the primary receptor for fibrinogen on the platelet. Although GPIIb–IIIa has a predominant role in platelet aggregation, it has been suggested that under conditions of high shear blood flow both GPIb and GPIIb–IIIa are involved in platelet adhesion. The importance of GPIIb–IIIa in aggregation was first documented from studies of Glanzmann's thrombasthenic patients, whose platelets lack or have dysfunctional GPIIb–IIIa receptors. These platelets have absent aggregation to all agonists tested, as functional GPIIb–IIIa is the final common pathway to platelet aggregation. Patients diagnosed with Glanzmann's thrombasthenia have a normal platelet size distribution and a normal platelet count.

GPIIb–IIIa is a calcium-dependent heterodimer with approximately 40,000–80,000 copies per platelet. Although the majority of GPIIb–IIIa is located on the outer membrane surface, studies have demonstrated that there are internal GPIIb–IIIa molecules located on the α granules and on the membrane network of the open canalicular system that become expressed upon activation of the platelet. The extent of expression of these internal GPIIb–IIIa molecules and the final surface density is dependent upon the strength of the platelet agonist. In addition to mediating platelet aggregation responses and perhaps to some extent adhesion, GPIIb–IIIa is also critical to clot retraction, a process that consolidates the formed clot at the wound site. Through GPIIb–IIIa and platelet cytoskeletal protein interaction, the platelets with bound fibrinogen and fibrin contract along the fibrin strands to consolidate the formed clot.

GPIV

GPIV (CD36 or GPIIIb) is a membrane glycoprotein widely distributed on human cells. On human platelets, CD36 has been proposed to mediate the initial events of platelet adhesion to collagen and to serve as one of the receptors for thrombospondin (TS-1). CD36 has been implicated in clinical disorders. For example, CD36 expression is increased in myeloproliferative disorders. CD36 is also the binding site on endothelial cells for erythrocytes infected with *Plasmodium falciparum* malaria. There have been reports that some anti-CD36 antibodies and CD36 ligands induce activation of platelets. This activation is apparently independent of the FcγRII receptor and may involve the interaction of cytosolic protein kinases of the src family with the CD36 cytoplasmic tail.

CD9

CD9, a membrane surface protein expressed on platelets as well as many other cell types, belongs to a newly described family of proteins, the tetraspanins. Based on their primary structures, the tetraspanins are predicted to be single polypeptide chains with four highly hydrophobic putative transmembrane regions and two extracellular loops. Results from early studies using a platelet model system are consistent with roles for CD9 in platelet signal transduction. Platelets are activated when an anti-CD9 monoclonal antibody such as mAb7 is used to crosslink CD9 with the low-affinity Fc receptor (FcγRII; CD32), a potent signal-transducing membrane protein in platelets. Mab7-induced platelet activation is accompanied by increases in phosphatidylinositol 4,5-bisphosphate turnover, Ca^{2+} flux, and protein phosphorylation expected of a phospholipase C-mediated activation. For this reason, anti-CD9 mAbs have been used as a thrombin-like agonist for platelet activation studies.

Slupsky *et al.* (1997) demonstrated a measurable platelet activation response through CD9 when the contribution of signaling via CD32 was either reduced by blocking antibodies or by degranulation of platelets. Results from studies using platelets also suggested roles for CD9 in aggregation and adhesion. Molecular associations of CD9 with GPIIb–IIIa and with GPIb have also been demonstrated. Furthermore, it has been shown that exposure of platelets to immobilized anti-CD9 F(ab')$_2$ increases platelet activation and tyrosine phosphorylation of p72syk. Although the exact function of CD9 in platelets is not understood, it is possible that CD9 may facilitate platelet

TABLE 1.1

Major Platelet Membrane Surface Proteins and Their Ligands

Platelet membrane protein	Proposed receptor functions
GPIa–IIa (VLA-3, α2β1)	Collagen
GPIb–IX-V	vWF and thrombin
GPIc–IIa (VLA-5, α5β1)	Fibronectin
GPIIb–IIIa	Fibrinogen, vWF, fibronectin, vitronectin
VnR (αvβ3)	Fibrinogen, vWF, fibronectin, vitronectin
GPIV	Thrombospondin, collagen
GPVI	Collagen
CD9	Fibronectin

adhesion and spreading through its association with GPIIb–IIIa and other major adhesive molecules.

A summary of the major platelet proteins and their proposed function is provided in Table 1.1.

EVENTS ASSOCIATED WITH PLATELET AGGREGATION

GPIIb–IIIa Signaling

GPIIb–IIIa is involved in inside–out and outside–in (i.e., bidirectional) signaling across the platelet plasma membrane (Figure 1.4). The interaction of agonists with their receptors initiates activation of intracellular pathways (shown as a black box), leading to activation of the GPIIb–IIIa receptor that permits binding of fibrinogen and von Willebrand factor. The binding of these ligands bridges adjacent activated platelets, leading to formation of the platelet aggregate. In addition, binding of ligands to the receptor causes additional changes in the conformation of the GPIIb–IIIa receptor as measured by conformation-sensitive antibodies (e.g., the mAb D3) as well as phosphorylation of several proteins that may be involved in

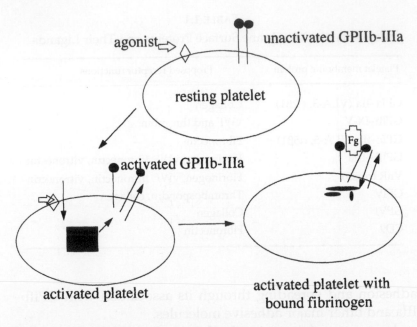

Figure 1.4. GPIIb–IIIa bidirectional signaling. GPIIb–IIIa expressed on the resting intact platelet does not bind plasma fibrinogen. Upon platelet activation, signaling molecules participate to induce activation of the GPIIb–IIIa receptor. The activated high-affinity receptor binds fibrinogen (or vWF under high shear conditions), and mechanochemical signals are transmitted across the membrane to mediate post-receptor occupancy events such as clot retraction.

signaling pathways that regulate platelet function. Therefore, signals are generated through GPIIb–IIIa from inside–out and from outside–in to facilitate aggregate formation.

GPIIb–IIIa activation occurs through generation or release of such soluble agonists as thrombin, thromboxane A$_2$, ADP, or serotonin. These agents do not act directly on GPIIb–IIIa but bind to platelet receptors, which leads to engagement of classical signal transduction pathways. G proteins and tyrosine kinases usually transduce these signals, which results in activation of phospholipase C (PLC), changes in cytosolic calcium, and activation of cellular protein kinases.

It has been proposed that "inside–out" signaling is mediated by the cytoplasmic domains of GPIIb–IIIa in response to intracellular signaling events initiated by platelet agonists. Recent work has implicated intracellular proteins such as calreticulin and serine/threonine kinases, and members of the small GTP-ase family such as R-Ras and RhoA. GPIIb–IIIa is constitutively in a low-affinity state and cannot bind soluble fibrinogen or vWF until it is activated via signal transduction mechanisms into a high-affinity ligand binding conformation. Inhibitors of these signal transduction pathways can affect agonist-induced ligand affinity of integrins. In response to vWF or fibrinogen binding to GPIIb–IIIa, additional conformational changes occur in the activated integrin that can be measured by a specialized class of monoclonal antibodies (mAbs) called anti-LIBSs (ligand-induced binding sites). Anti-LIBSs such as D3 have been used to measure ligand binding to GPIIb–IIIa. In addition, these mAbs have been used to directly activate GPIIb–IIIa on platelets and in transfected cells, as well as to alter GPIIb–IIIa-mediated functions. Certain anti-LIBSs inhibit clot retraction, full-scale platelet aggregation, or reversible aggregation presumably by interrupting outside–in signaling events that occur upon receptor occupancy. The exact function that LIBSs have in regulation of integrin high-affinity states is still under investigation. There are reports that the GPIIIa cytoplasmic tail is phosphorylated predominantly on serine and threonine residues upon platelet activation and that phosphorylation of these sites may regulate exposure of LIBSs. Other intracellular elements bind to the cytoplasmic domains of integrins (Figure 1.4) to create an integrin activating complex including talin and α-actinin. Furthermore, it has been demonstrated that the GPI-IIa subunit is phosphorylated in response to thrombin-mediated platelet aggregation. The data suggest that tyrosine phosphorylation of the GPIIIa (beta 3) subunit of an integrin may be key in outside–in signaling by facilitating association of signaling molecules with the integrin complex.

Displacement of divalent cations during ligand binding may trigger conformational changes across the membrane. It is ap-

Conformational States of GPIIb-IIIa

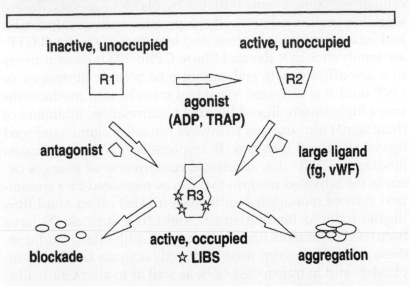

Figure 1.5. The conformational state of GPIIb–IIIa when it is in an inactive unoccupied state vs. an active occupied state. The inactive receptor (R1) represents the GPIIb–IIIa that is expressed on the resting platelet. After platelets are activated by agonist, the R1 conformer converts into an active receptor, R2. If large ligand is present, the ligand will bind, resulting in a further conformational change represented by R3. LIBSs represented as stars are expressed. Binding of fibrinogen or vWF and the crosslinking of adjacent platelets results in platelet aggregation. On the other hand, if resting platelets are treated with GPIIb–IIIa receptor antagonists, conformer R1 converts directly to R3. Because the small antagonists serve as competitive inhibitors of large ligand binding, blockade of aggregation occurs.

parent that GPIIb–IIIa is a dynamic protein regulated by the activation state of the platelet and is responsive to ligation by adhesive molecules. Understanding the mechanism of activation of this receptor will provide a means to regulate the activity of platelets in hemostasis and in thrombosis. Figure 1.5 demonstrates the dynamic conformational states of GPIIb–IIIa. The GPIIb–IIIa molecule is in a resting conformer (R1) that cannot bind soluble ligands such as fibrinogen or vWF. Upon platelet activation, a conformational change occurs in GPIIb–IIIa resulting in conformer R2. Ligand binds, generating an active ligand-occupied receptor, R3, that is recognized by anti-LIBS antibodies. The addition of an antagonist to resting platelets can cause direct transformation of R1 to R3 since these antagonists can bind to GPIIb–IIIa due to their small size without the requirement of R2 formation. Because these antagonists block the binding of fibrinogen and vWF, blockade of aggregation occurs.

Platelet Cytoskeleton

One of the major platelet proteins is actin. Many platelet responses require the force generated by contractile proteins. These responses include secretion, shape change, platelet aggregation, and clot retraction. The platelet cytoskeleton — which contains actin, a large number of actin-binding proteins, myosin, and myosin-associated proteins — is in a dynamic state. Resting platelets contain a microfilament (MF) network and a membrane web of short actin filaments (or submembrane filaments [SMFs]) that may be important in maintaining the structural integrity of the unactivated discoid platelet (Figure 1.6). The primary membrane surface component of this web is the GPIb–IX complex. Actin-binding protein, a 250-kilodalton dimer, provides a linkage of the platelet filaments to GPIb–IX. Upon platelet activation and a resultant increase in cytosolic calcium levels, an enzyme called calpain becomes activated and cleaves actin-binding protein, releasing this linkage and allowing increased mobility of platelet membrane components.

During platelet activation, the platelet undergoes a dramatic shape change, transforming from a discoid shape to an irregular rounded shape with extensive pseudopodia. This morphological change is associated with rearrangement of the platelet cytoskeleton including the microtubules (MTs) (Figure 1.7).

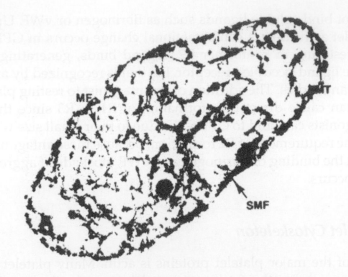

Figure 1.6. Microfilament network of a resting discoid platelet. MF, microfilament; SMF, submembranous filament; MT, microtubule.

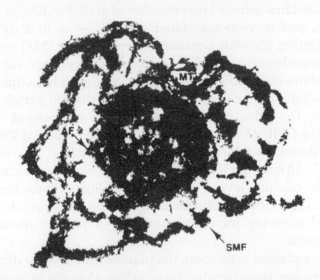

Figure 1.7. Cytoskeletal organization of an activated platelet. Note the dense core of filaments (contractile gel) as well as the pseudopodal cytoskeleton. MT, microtubule; AF, actin filament; SMF, submembranous filament.

The percentage of F- or filamentous actin increases from about 15% of the total platelet protein in the resting platelet to about 70% in the activated platelet. Platelet myosin also becomes phosphorylated and associates with the actin filaments. In addition, new linkages of the platelet cytoskeleton components with GPIIb–IIIa occurs along with localization of signaling molecules that facilitate platelet aggregation and clot retraction. Cytoskeleton rearrangements and protein composition can be studied by isolating both the membrane-associated cytoskeleton and core cytoskeleton from platelets using the Triton X-100 extraction technique. The core cytoskeleton can be isolated by low-speed centrifugation after solubilizing the platelets with a final 1% Triton X-100 containing buffer (see the Appendix). Membrane-associated cytoskeletons are isolated by centrifuging the low speed supernatant at 100,000g. The protein composition of these cytoskeletons is routinely analyzed by polyacrylamide–SDS gel electrophoresis.

Storage Granules and the Release Reaction

Platelets contain an extensive number of active molecules within their storage granules (Figure 1.8; see Table 1.2). The alpha (α) granules contain coagulant and adhesive proteins such as vWF, fibrinogen, fibronectin, vitronectin, TSP, and Factor V, as well as growth factors and inhibitors. The release of α-granule proteins PF4 and βTG has been used as an indicator of platelet release. P-selectin, an α-granule membrane component (also known as CD62P, GMP-140, and PADGEM), is translocated to the platelet surface as well as to the endothelial cell surface upon activation. This antigen has been shown to promote platelet–leukocyte interactions. The expression of P-selectin on the platelet surface has been used as a marker for platelet activation. It has also been postulated that raised levels of soluble P-selectin may be a marker for thrombotic disease.

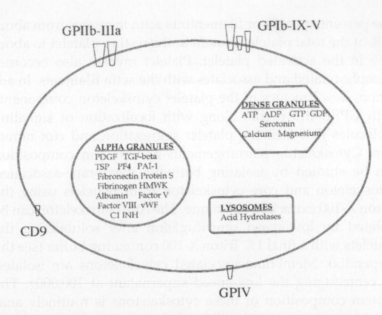

Figure 1.8. Schematic representation of platelet granule contents and major surface integral proteins.

TABLE 1.2

Platelet Granules and Their Contents

α-granules	Dense granules	Lysosomes
PDGF	ATP	Acid hydrolases
TGF-β	ADP	
CTAP III	GTP	
Platelet factor 4	GDP	
TSP	Serotonin	
Fibronectin	Calcium	
Fibrinogen	Magnesium	
Vitronectin		
von Willebrand factor		
Albumin		
Factors V and VIII		
Protein S		
PAI-1		
HMWK		
C1 INH		

PDGF = platelet derived growth factor; TGF-β = transforming growth factor-beta; CTAP III = connective tissue activating peptide; PAI-1 = plasminogen activator inhibitor-1; HMWK = high-molecular-weight kininogen; C1 INH = C1 inhibitor; TSP = thrombospondin.

The dense bodies store serotonin, calcium, adenine and guanine nucleotides, divalent cations, and P_i. Lysosomes contain such hydrolytic enzymes as acid hydrolases, cathepsin, and heparitinase. In addition, peroxisomes have been identified that contain catalase. Generally, the release of the contents of these storage organelles reflects the stimulus intensity to which platelets are exposed.

PLATELET AGONISTS

Adenosine Diphosphate (ADP)

Addition of ADP to platelets causes shape change and exposure of the fibrinogen binding site on GPIIb–IIIa, fibrinogen binding, and platelet aggregation (Figure 1.9A). ADP induces an increase in the cytoplasmic free calcium level by release from internal stores as well as by Ca^{2+} influx. ADP inhibits stimulated adenyl cyclase and also causes reorganization of the cytoskeleton of platelets. ADP is a weak activator of phospholipase C. It has been proposed that ADP may activate platelets through a G protein identified as G_{i2}. Existing evidence suggests that there may be two receptors on the platelet for ADP: one responsible for mediating platelet aggregation that is coupled to the G protein and the other responsible for calcium influx and shape change.

Epinephrine

Epinephrine is unique among the platelet agonists because it causes platelet aggregation and release but not shape change (Figure 1.9B). Activation of phospholipase C by epinephrine is dependent upon the formation of thromboxane A_2 and is inhibited by treatment with aspirin. However, it appears that thromboxane A_2 formation is not absolutely required for epinephrine-induced platelet aggregation, as GPIIb–IIIa activation and fibrinogen binding can still occur in aspirin-inhibited platelets when exposed to epinephrine. Other mechanisms of action include the possible influence of epinephrine on the rate

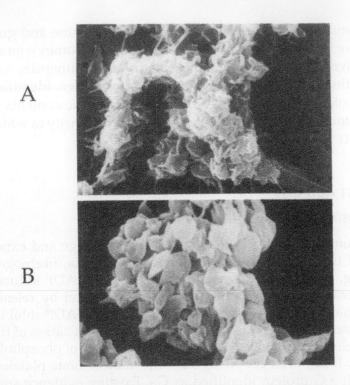

Figure 1.9. Scanning electron micrographs of (A) ADP- and (B) epi-nephrine-induced aggregates. Note the smooth appearance of the platelets in the epinephrine-induced aggregate as opposed to the amorphous mass of platelets with extended psuedopods in the ADP-induced aggregate.

of Na^+/H^+ exchange across the platelet membrane, which may affect PLA_2 activation. Platelet responses to epinephrine are highly variable.

Collagen

Collagen is an insoluble protein that induces activation, adhe-sion, granule release, and aggregation of platelets. Platelet ag-gregation induced by low-dose collagen is highly dependent upon the release reaction and is very sensitive to aspirin inhibition. Several candidates have been selected as collagen receptors. The two most prominent proteins are glycoprotein Ia/IIa (integrin $\alpha_2\beta_1$) and GPVI. GPIV has also been identified as having collagen-binding activity. As with other stimulants like ADP and thrombin, collagen activation of platelets results in an increase in cytosolic calcium concentration and generation and turnover of the phosphoinositides. Perfusion chamber experiments that mimic the conditions of blood flow show that platelet adhesion to collagen is dependent upon von Willebrand factor interaction with platelet GPIb. vWF binds to the exposed collagen surface, and platelets adhere to the immobilized vWF via GPIb. Adherent platelets also interact with the immobilized collagen, probably through GPIa/IIa and GPVI, resulting in platelet activation and recruitment of nearby platelets to form an aggregate.

Thrombin

Thrombin is the most potent platelet agonist. The thrombin receptor is a 425-amino acid integral protein that spans the membrane seven times. It has a large amino-terminal extracellular region that undergoes thrombin cleavage at amino acid Arg-41. Cleavage of this amino terminus region creates a newly exposed amino terminus that ligates to another region of the receptor as a tethered ligand. Thrombin activation of platelets is mediated by the activation of phospholipase C (PLC) and phospholipase A_2 (PLA$_2$). PLC catalyzes the breakdown of plasma membrane inositol phospholipids, resulting in generation of 1,2-diacylglycerol (DAG) and 1,4,5-inositol triphosphate (IP$_3$). DAG activates protein kinase C and IP$_3$ induces mobilization of calcium from intracellular stores. PLA$_2$ mediates the generation of arachidonic acid, which can be converted into the potent platelet agonist thromboxane A_2 (TXA$_2$). TXA$_2$ can diffuse across the plasma membrane and bind to TXA$_2$ receptors, thus potentiating the activation response of platelets. Other events include protein kinase C activation, phosphatidylinositol 3-kinase activation, actin polymerization, myosin light chain phosphorylation, activation of tyrosine kinases, and activation of GPIIb–IIIa.

Antibody Activation of Platelets

The activation of platelets by monoclonal antibodies can be mediated by direct interaction of an antibody with its antigen, by the crosslinking of antigens, or by the crosslinking of antigen to the Fc receptor. It was originally thought that the average number of FcγRII receptors on the platelet surface was approximately 1000. However, the use of anti-FcγRII Fab fragments rather than the bivalent antibody suggests approximately 3000 FcγRII receptors per platelet. The role of the Fc receptor in platelet activation is not fully understood; however, FcγRII polymorphism should be considered due to the fact that the antibody Fc domain has a higher affinity for the Arg form of FcγRIIa than for the His form. Platelet activation via Fc receptor and antigen crosslinking is generally very potent and similar to that induced by thrombin.

COAGULATION

Activation of the coagulation cascade occurs by exposure of de-endothelialized vessel wall and release of tissue factor. Thrombin is generated, which activates platelets and catalyzes generation and formation of fibrin from fibrinogen. Fibrin generation is considered key in stabilization of the platelet thrombus and is resistant to removal at high shear rates. Therefore, platelets and the coagulation cascade are intimately interrelated in the pathogenesis of thrombosis. Platelet activation results in increased procoagulant activity and formation of factor Va and VIIIa binding sites through increased exposure of phosphatidylserine at the platelet outer membrane surface. The culmination of these events is enhancement of thrombin formation. Furthermore, platelet activation results in formation of microparticles that also contain clot-promoting activity. The importance of the expression of this procoagulant activity is underscored by identification of a bleeding disorder called Scott syndrome, which is due to a deficiency in the generation of platelet procoagulant activity and in the shedding of platelet microparticles.

REFERENCES

Bachelot, C., Saffroy, R., Gandrille, S., Aiach, M., and Rendu, F. (1995). Role of FcγRIIA gene polymorphism in human platelet activation by monoclonal antibodies. *Thromb. Haemost.* **74**, 1557–1563.

Bolin, R. B., Okumura, T., and Jamieson, G. A. (1977). Changes in the distribution of platelet membrane glycoproteins in patients with myeloproliferative disorders. *Am. J. Hematol.* **3**, 63–71.

Calvete, J. J., Mann, K., Schafer, W., Fernandez-Lafuente, R., and Guisan, J. M. (1994). Proteolytic degradation of the RGD-binding and non-RGD-binding conformers of human platelet integrin glycoprotein IIb/IIIa: clues for identification of regions involved in the receptor's activation. *Biochem. J.* **298**, 1–7.

Clemetson, K. J. (1989). Platelet GPIb–V–IX complex. *Thromb. Haemost.* **78**, 266–270.

D'Souza, S. E., Ginsberg, M. H., Burke, T. A., Lam, S. C., and Plow, E. F. (1988). Localization of an Arg–Gly–Asp recognition site within an integrin adhesion receptor. *Science* **242**, 91–93.

D'Souza, S. E., Ginsberg, M. H., Burke, T. A., and Plow, E. F. (1990). The ligand binding site of the platelet integrin receptor GPIIb–IIIa is proximal to the second calcium binding domain of its subunit. *J. Biol. Chem.* **265**, 3440–3446.

D'Souza, S. E., Haas, T. A., Piotrovicz, R. S., Byers-Ward, V., McGrath, D. E., Soule, H. R., Cierniewski, C., Plow, E. F., and Smith, J. W. (1994). Ligand and cation binding are dual functions of a discrete segment of the β3 integrin: cation displacement is involved in ligand binding. *Cell* **70**, 659–667.

Fijnheer, R., Frijns, C. J. M., Korteweg, J., Rommes, H., Peters, J. H., and Sixma, J. J. (1997). The origin of P-selectin as a circulating plasma protein. *Thromb. Haemost.* **77**, 1081–1085.

Finkle, C. D., St. Pierre, A., Leblond, L., Deschenes, I., DiMaio, J., and Winocour, P. D. (1998). BCH-2763, a novel potent thrombin inhibitor, is an effective antithrombotic agent in rodent models or arterial and venous thrombosis — comparisons with heparin, r-hirudin, hirulog, inogatran, and argatroban. *Thromb. Haemost.* **79**, 431–438.

Fox, J. E. B. (1985). Identification of actin-binding protein linking the membrane cytoskeleton to glycoproteins on platelet plasma membranes. *J. Biol. Chem.* **260**, 11970–11977.

Frelinger, A. L., Lam, S. C. T., Plow, E. F., Smith, M. A., Loftus, J. C., and Ginsberg, M. H. (1988). Occupancy of an adhesive glycoprotein receptor modulates expression of an antigenic site involved in cell adhesion. *J. Biol. Chem.* **263**, 12397–12402.

Gachet, C., Cattaneo, M., Ohlmann, P., Lecchi, A., Hechler, B., Chevalier, J., Cassel, D., Mannucci, P. M., and Cazenave, J. P. (1995). Purinoreceptors on blood platelets: further pharmacological and clinical evidence to suggest the presence of two ADP receptors. *Br. J. Haematol.* **91**, 434–444.

Gachet, C., Hechler, B., Leon, C., Vial, C., Leray, C., Ohlmann, P., and Cazenave, J.-P. (1997). Activation of ADP receptors and platelet function. *Thromb. Haemost.* **78**, 271–275.

Ginsberg, M. H., Lightsey, A., Kunicki, T. J., Kaufman, A., Marguerie, G. A., and Plow, E. F. (1986). Divalent cation regulation of the surface orientation of platelet membrane glycoprotein IIb: correlation with fibrinogen binding function and definition of a novel variant of Glanzmann's thrombasthenia. *J. Clin. Invest.* **78**, 1103–1111.

Gulino, D., Boudignon, C., Zhang, L., Concord, E., Rabiet, M. J., and Marguerie, G. (1992). Ca^{2+} binding properties of the platelet glycoprotein IIb ligand-interaction domain. *J. Biol. Chem.* **267**, 1001–1007.

Haimovich, B., Lipfert, L., Brugge, J., and Shattil, S. (1993). Tyrosine phosphorylation and cytoskeletal reorganization in platelets triggered by interaction of integrin receptors with their immobilized ligands. *J. Biol. Chem.* **268**, 15868–15877.

Harrison, P., and Cramer, E. M. (1993). Platelet α-granules. *Blood Rev.* **7**, 52–62.

Heath-Mondoro, T., Wall, D. C., White, M. M., and Jennings, L. K. (1996). Selective induction of a glycoprotein IIIa ligand-induced binding site by fibrinogen and von Willebrand factor. *Blood* **88**, 3824–3830.

Horwitz, A., Duggan, K., Buck, C., Beckerle, M., and Burridge, K. (1986). Interaction of plasma fibrinogen receptor with talin, a transmembrane linkage. *Nature (London)* **320**, 531–533.

Hotchin, N. A., and Hall, A. (1995). The assembly of integrin adhesion complexes requires both extracellular matrix and intracellular Rho/rac GTPases. *J. Cell. Biol.* **131**, 1857–1865.

Huang, M. M., Bolen, J. B., Barnwell, J. W., Shattil, S. J., and Brugge, J. S. (1991). Membrane glycoprotein IV (CD36) is physically associated with the Fyn, Lyn, and Yes protein-tyrosine kinases in human platelets. *Proc. Natl. Acad. Sci. U.S.A.* **88**, 7844–7848.

Irie, A., Kamata, T., Puzon-McLaughlin, W., and Takada, Y. (1995). Critical amino acid residues for ligand binding are clustered in a predicted beta-turn of the third N terminal repeat in the integrin alpha 4 and alpha 5 subunits. *EMBO J.* **14**, 5550–5556.

Jennings, L. K., Fox, J. E. B., Edwards, H. H., and Phillips, D. R. (1981). Changes in the cytoskeletal structure of human platelets following thrombin activation. *J. Biol. Chem.* **256**, 6927–6934.

Jennings, L. K., White, M. M., and Mandrell, T. D. (1995). Interspecies comparison of platelet aggregation, LIBS expression and clot retraction: observed differences in GPIIb–IIIa functional activity. *Thromb. Haemost.* **74**, 1551–1556.

King, M., McDermott, P., and Schreiber, A. D. (1990). Characterization of the Fcg receptor on human platelets. *Cell Immunol.* **128**, 462–479.

Kobe, B., and Deisenhofer, J. (1994). The leucine-rich repeat: a versatile binding motif. *TIBS* **19**, 415–421.

Kouns, W. C., Wall, C. D., White, M. M., Fox, C. F., and Jennings, L. K. (1990). A conformation-dependent epitope of human platelet glycoprotein IIIa. *J. Biol. Chem.* **265**, 20593–20601.

Law, D. A., Nannizzi-Alaimo, L., and Phillips, D. R. (1996). Outside–in integrin signal transduction. *J. Biol. Chem.* **271**, 10811–10815.

Leung, L. L. K., Li, W.-X., McGregor, J. L., Albrecht, G., and Howard, R. J. (1992). CD36 peptides enhances or inhibits CD36-thrombospondin binding: a two-step process of ligand-receptor interaction. *J. Biol. Chem.* **267**, 18244–18250.

Loftus, J. C., O'Toole, T. E., Plow, E. F., Glass, A., Frelinger, A. L., and Ginsberg, M. H. (1990). A b3 integrin mutation abolishes ligand binding and alters divalent cation-dependent conformation. *Science* **249**, 915–918.

Moroi, M., Jung, S. M., Okumqa, M., and Shinmyozu, K. (1989). A patient with platelets deficient in glycoprotein VI that lack both collagen-induced aggregation and adhesion. *J. Clin. Invest.* **84**, 1440–1445.

Nieuwenhuis, H. K., Akkerman, J. W. N., Houdijk, W. P. M., and Sioxma, J. J. (1985). Human blood platelets showing no response to collagen fail to express glycoprotein Ia. *Nature* **264**, 6017–6020.

Ockenhouse, C. F., Tandon, N. N., Magowan, C., Jamieson, G. A., and Chulay, J. (1989). Identification of a platelet membrane glycoprotein as a *falciparum* malaria sequestration receptor. *Science* **243**, 1469–1471.

Ohlmann, P., Laugwitz, K. L., Spicher, K., Nurnberg, K., Schultz, G., Cazenave, J. P., and Gachet, C. (1995). The human platelet ADP receptor activates Gi2 proteins. *Biochem. J.* **312**, 775–779.

Otey, C. A., Vasquez, G. B., Burridge, K., and Erickson, B. W. (1993). Mapping of the alpha-actinin binding site within the beta 1 integrin cytoplasmic domain. *J. Biol. Chem.* **268**, 21193–21197.

O'Toole, T. E., Katagiri, Y., Faull, R. J., Peter, K., Tamura, R., Quaranta, V., Loftus, J. C., Shattil, S. J., and Ginsberg, M. H. (1994). Integrin cytoplasmic domains mediate inside–out signal transduction. *J. Cell. Biol.* **124**, 1047–1059.

Parise, L. V., Criss, A. B., Nannizzi, L., and Wordell, M. R. (1990). Glycoprotein IIIa is phosphorylated in intact human platelets. *Blood* **75**, 2363–2368.

Rosenfeld, S. L., and Anderson, C. L. (1989). Fc receptors of human platelets. *In* "Platelet Immunobiology, Molecular and Clinical Aspects" (T. J. Kunicki and J. N. George, eds.), pp. 337–353. Lippincott, Philadelphia.

Shattil, S. J., and Brass, L. F. (1987). Induction of fibrinogen receptor on human platelets by intracellular mediators. *J. Biol. Chem.* **262**, 992–1000.

Sims, P. J., Wiedmer, T., Esmon, C. T., Weiss, H. J., and Shattil, S. J. (1989). Regulatory control complement on blood platelets: modulation off platelet precoagulant responses by a membrane inhibitor of the C5b–9 complex. *J. Biol. Chem.* **264**, 17049–17057.

Slupsky, J. R., Cawley, J. C., Kaplan, C., and Zuzel, M. (1997). Analysis of CD9, CD32, and p67 signalling: use of degranulated platelets indicates direct involvement of CD9 and p67 in integrin activation. *Br. J. Haematol.* **96**, 275–286.

Tandon, N. N., Kralisz, U., and Jamieson, G. A. (1989). Identification of GPIV (CD36) as a primary receptor for platelet collagen adhesion. *J. Biol. Chem.* **264**, 7576–7583.

Van Willigen, G., Hers, I., Gorter, G., and Akkerman, J.-W. N. (1996). Exposure of ligand binding sites on platelet integrin aIIbb3 by phosphorylation of the b3 subunit. *Biochem. J.* **314**, 769–779.

Vu, T. K., Hung, D. T., Wheaton, V. I., and Coughlin, S. R. (1991). Molecular cloning of a functional thrombin receptor reveals a novel proteolytic mechanism of receptor activation. *Cell* **64**, 1057–1068.

Weiss, H. J., Vivic, W. J., Lages, B. A., and Rogers, J. (1979). Isolated deficiency of platelet procoagulation activity. *JAMA* **67**, 206–213.

Williams, M. J., Hughes, P. E., O'Toole, T. E., and Ginsberg, M. H. (1994). The inner world of cell adhesion: integrin cytoplasmic domains. *Trends Cell. Biol.* **4**, 109–112.

Wu, Y. P., van Breugel, H. H, F, I., Lanhof, H., Wise, R. J., Handin, R. I., de Groot, P. G., and Sixma, J. J. (1996). Platelet adhesion to multimeric and dimeric von Willebrand factor and to collagen type III preincubated with von Willebrand factor. *Arterioscler. Thromb. Vasc. Biol.* **16**, 611–620.

Zhang, L., and Plow, E. (1996). Overlapping but not identical sites are involved in the recognition of C3bi neutrophil inhibitory factor and adhesive ligands by the alphaMbeta2 integrin. *J. Biol. Chem.* **271**, 18211–18216.

LABORATORY EVALUATION OF PLATELET FUNCTION

INTRODUCTION

Platelet aggregation has been the method of choice for assessing platelet function since the early 1970s. The instrumentation for measurement of platelet aggregation has changed very little during that time, although the advent of lumiaggregometry has made the assessment of platelet nucleotide content and release considerably less difficult. When asked to evaluate a patient for a platelet function defect, it is recommended that you examine a peripheral smear and conduct a platelet count in addition to measuring platelet aggregation. Lumiaggregometry or another assay of granule release can be used to distinguish storage pool disease (SPD) and release defects. Flow cytometric analysis of GPIb and GPIIb–IIIa surface density can be used to aid in the diagnosis of Bernard–Soulier syndrome or Glanzmann's thrombasthenia. When performing aggregation, as with any other laboratory procedure, it is essential to control variables that might affect the test results. Some of the variables encountered in performing platelet aggregation will be covered in the following sections.

VARIABLES

Anticoagulants

Citrate. Both 0.1 and 0.129 M citrate (buffered and nonbuffered) at a ratio of 9 parts blood to 1 part anticoagulant are traditionally used for platelet aggregation testing. A correction formula for abnormal hematocrits can be found in the section on Drawing and Processing Blood for Platelet Aggregation. For typical agonists such as ADP, collagen, and epinephrine, at moderate concentrations there is little or no difference seen in the test results between blood drawn into 0.1 or 0.129 M citrate. A discrepancy may be seen when testing in the lowest concentration ranges of agonists such as ADP, where higher concentrations of citrate can result in lower aggregation responses. This can be exacerbated if the patient or donor has a high hematocrit that is not corrected for, thus increasing the citrate-to-plasma ratio. Also, if a situation arises where the specimen will sit on the bench for more than 1–2 hr, it is definitely preferable to use a buffered anticoagulant to help maintain the pH of the PRP. Figure 2.1 demonstrates the aggregation response to ADP in three anticoagulants: 0.1 M buffered citrate, 0.129 M sodium citrate, and ACD (acid–citrate–dextrose). The latter is only used for preparation of washed platelets and not for aggregation assays (see below). The anticoagulant of choice for aggregation studies is 0.1 M buffered citrate.

BEWARE: Manufacturers of vacutainer tubes use the same blue color stopper for all citrate tubes regardless of concentration or buffering capacity.

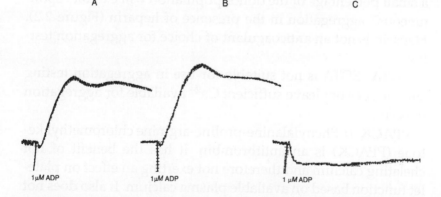

Figure 2.1. ADP response in (A) 0.1 M buffered citrate, (B) 0.129 M sodium citrate, and (C) acid–citrate–dextrose, ACD. These assays were performed within 90 min of blood sample collection.

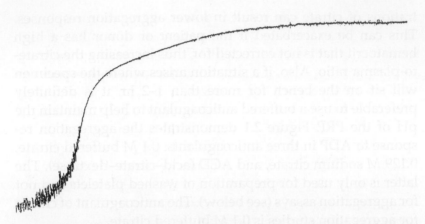

Figure 2.2. This spontaneous aggregation was obtained in the PRP from a dialysis patient post-dialysis. The patient was found to have a heparin-induced antibody, and the "spontaneous" aggregation observed was due to residual heparin present in the plasma sample.

Heparin. Heparin can be used for platelet testing, but in many donors the PRP platelet count will be significantly lower when drawn into heparin as compared to citrate. In addition, a small percentage of the donor population will exhibit "spontaneous" aggregation in the presence of heparin (Figure 2.2). Heparin is not an anticoagulant of choice for aggregation testing.

EDTA. EDTA is not suitable for use in aggregation testing since it does not leave sufficient Ca^{2+} available for aggregation to occur.

PPACK. D-Phenylalanine-proline-arginine chloromethyl ketone (PPACK) is an antithrombin. It has the benefit of not chelating calcium and therefore not exerting an effect on platelet function based on available plasma calcium. It also does not have the proaggregating effects of heparin. It has been demonstrated that the plasma free calcium concentration has a direct effect on the IC_{50} of inhibition of platelet aggregation by some

GPIIb–IIIa antagonists currently being evaluated in the treatment of acute coronary syndromes. In order to more accurately reflect the *in vivo* situation, PPACK-anticoagulated blood is being used in the evaluation of the extent of inhibition of aggregation by these drugs. Unfortunately, PPACK is very expensive relative to the other anticoagulants listed, and its anticoagulant effect is finite. Using a final concentration of 1.6 mg per 10 ml whole blood (0.3 mM final concentration), one can anticipate that the specimen will not clot for several hours. To provide an anticoagulant effect for more than 24 hr, we recommend a final concentration of 12.8 mg per 10 ml (2.4 mM final) whole blood.

ACD. Because this anticoagulant brings the pH of the PRP to 6.5, it is not suitable for use in aggregation experiments (Figure 2.3). It is an excellent choice if the purpose is to obtain platelets for washing or gel-filtration. Use at a ratio of 6 parts blood to 1 part anticoagulant.

ACD-A. This formulation of ACD keeps the pH of the PRP at 7.2 and has been reported as acceptable for aggregation testing.

At this point, it should be noted that anticoagulants such as PPACK and heparin cannot be used in aggregation testing if simultaneous measurements of release are being made (Figure 2.4). The release reaction seen in citrated blood, and which is used in the diagnosis of SPD and release defects, is not seen with the agonist ADP in blood anticoagulated with these antithrombins. Aggregation responses to ADP are generally lower in PPACK- or heparin-anticoagulated blood compared to citrate-anticoagulated blood (see Figure 3.1, Chapter 3).

pH

Platelet aggregation should be performed at pH 7.2–7.4. At lower pH, platelet responsiveness is diminished to the point of total inhibition of aggregation at pH 6.5. On the other hand, as the pH rises, so does the response of the platelets. At pH 8.1

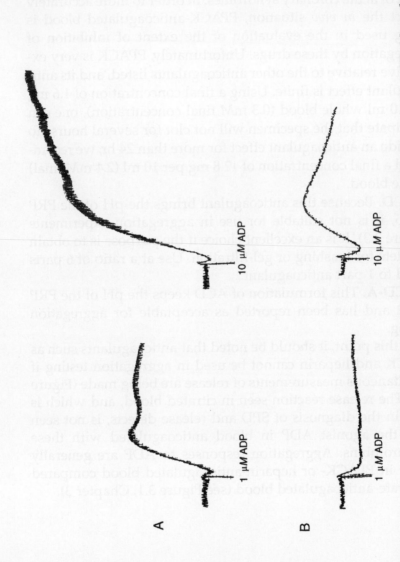

Figure 2.3. Response to 1 μM ADP and 10 μM ADP in (A) 0.1 M buffered citrate and (B) ACD. Note the profound inhibition of aggregation from the blood collected in ACD.

NORMAL – IN PPACK

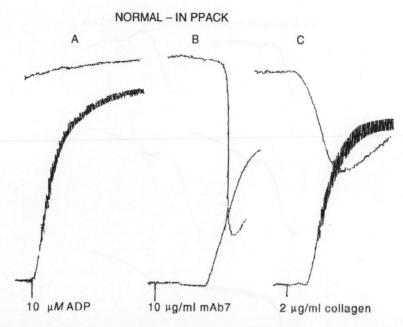

Figure 2.4. Aggregation and release tracings from a normal donor drawn into PPACK anticoagulant reflecting responses to typical agonists used in the laboratory. Deflection in baseline tracing indicates point at which agonist is added. Aggregation response is lower tracing; release response is upper tracing. Note absence of release in response to ADP stimulation compared to that obtained in citrate anticoagulant (see Figure 3.1, Chapter 3).

and higher, spontaneous aggregation can occur. It is best to store platelets in tubes that are capped and with very little air space relative to PRP volume (Figure 2.5A vs. 2.5B).

Temperature

Platelet aggregation should be performed at 37°C to mimic the *in vivo* situation. Studies have shown that exposure of platelets to cold temperatures can lead to a spontaneous aggregation response upon rewarming and stirring of the platelet suspen-

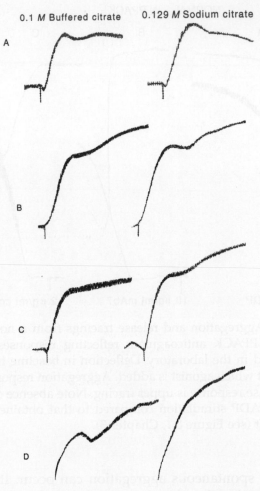

0.1 *M* Buffered citrate   0.129 *M* Sodium citrate

A

B

C

D

Figure 2.5. Effect of platelet handling on platelet aggregation responses. (A) Platelets were prepared and assayed as recommended in the text. (B) Platelets were prepared and then exposed to conditions under which the pH changed from ~7.4 in both anticoagulants to ~8.0 in buffered citrate, and to ~8.5 in 0.129 *M* citrate. Aggregation response to 1 μM ADP was measured. Note the increase in aggregation response due to an increase in the plasma pH. (C) PRP was prepared and then placed in an ice water bath for 15 min before aggregation testing with 1 μM ADP. Note the increased response over that of correct handling (A). (D) Platelets were prepared and stored at 4°C for 2 hr (i.e., refrigeration). Note the lack of a baseline tracing and the absence of agonist addition. Before either of these could be accomplished, spontaneous aggregation occurred in PRP prepared in both anticoagulants.

sion. Furthermore, when the pre-chilled platelets have been incubated at warmer temperatures for longer periods, the agonist-induced aggregation response is higher compared to a control platelet sample. Thus, incorrect storage of platelets can lead to erroneous values for aggregation response (Figure 2.5A vs. 2.5C and 2.5D).

Glass vs. Plastic Tubes

Platelet preparation should always be carried in either plastic or siliconized glass tubes. Uncoated glass will cause platelet activation. We have found polypropylene to be superior to polystyrene or polycarbonate when used for platelet preparation or storage.

Aggregometer Stir Speed

Platelets must come into contact with each other to be able to aggregate (the propinquity effect). If agonists are added to nonstirred platelets, they will become activated and perhaps form microaggregates, but they will not form visible aggregates unless agitated (Figure 2.6). The optimal stir speed for any instrument must be determined based on the height of the PRP column, the diameter of the cuvette, and the stir bar used. Guidelines for settings are recommended by the aggregometer manufacturer.

Platelet Count

PRP should be adjusted to a constant platelet count using autologous PPP to ensure the least assay-to-assay variability. Platelet aggregation responses can be affected by the platelet count of the PRP, although the count must be very low before a major difference in aggregation results will be seen. Measurements of nucleotide release are significantly affected by plate-

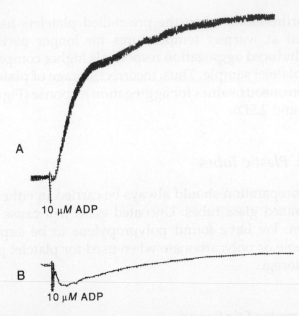

Figure 2.6. Platelet aggregation response to 10 μM ADP in the presence (A) or absence (B) of stirring. To obtain an aggregation response, the platelets not only must be activated but also come into contact with each other. If an aggregation pattern such as (B) is observed, check for the presence of a stir bar and that it is stirring in the cuvette before reporting "0%" aggregation.

let number (Figures 2.7 and 2.8). If it is necessary to perform release studies at a count lower than that used to establish a normal range, the results can be corrected. Usually, platelet aggregation studies are performed at counts of 250,000–300,000/mm^3.

Fibrinogen

Platelet aggregation requires the presence of fibrinogen. Very low levels of fibrinogen or fibrinogen with an abnormal structure can inhibit platelet aggregation.

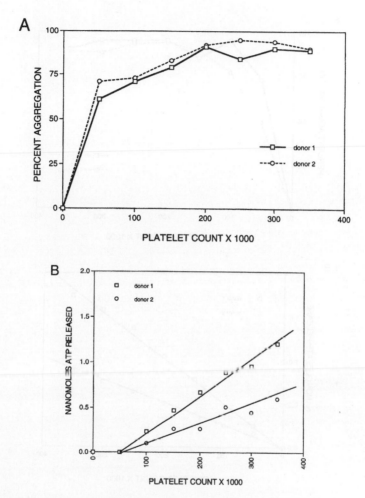

Figure 2.7. Platelet aggregation (A) and release (B) as a function of platelet count in response to 10 μM ADP.

Hemolysis

Nucleotides released from lysed red blood cells can affect platelet function, making them refractory to the addition of ADP.

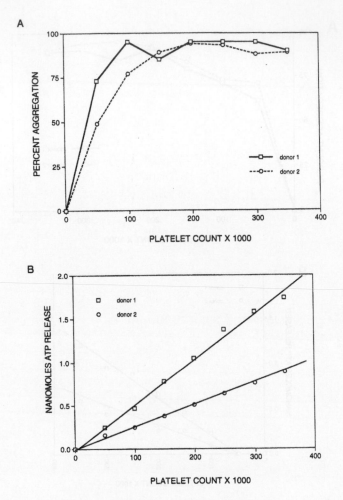

Figure 2.8. Platelet aggregation (A) and release (B) as a function of platelet count in response to 20 μg/ml mAb7.

Red Blood Cell Contamination

The presence of red blood cells in PRP can inhibit the ability of the aggregometer to measure platelet aggregates and therefore can cause a decrease in percent aggregation.

Lipemia

The presence of lipids in the PRP can affect aggregation by interfering with the light transmission readings.

Venipuncture

Blood should be obtained using a 19–20 gauge needle and plastic syringe. We have not found a problem with using a single syringe, as opposed to the two-syringe technique, as long as the venipuncture is clean with no necessity for probing to find a vein. It is unclear whether vacutainer collection of blood is suitable for platelet function measurements. We have some data that indicate increased responsiveness to low-dose ADP in vacutainer as opposed to syringe PRP. Until vacutainers are proven to not increase the activation state of platelets or to affect pharmacodynamic measurements, we recommend the use of a syringe.

DRAWING AND PROCESSING BLOOD FOR PLATELET AGGREGATION

Method

1. Blood for platelet function studies should be drawn through a 19- or 20-gauge butterfly needle into a plastic syringe and added gently down the side of 15 ml plastic tubes containing 1 ml of buffered citrate anticoagulant.

NOTE: In the case of normal hematocrits, a ratio of 1 part anticoagulant to 9 parts blood is optimal.

The procedure for processing of blood to obtain PRP and PPP is outlined in Figure 2.9.

If the hematocrit is out of normal range, the following formula should be used to correct for the amount of blood added to 1 ml of anticoagulant so that the correct citrate:plasma ratio is maintained:

$$5/(1 - 0.\text{Hct}) = \text{amount of whole blood to add}$$
$$\text{to 1 ml anticoagulant}$$

Example: if the hematocrit is 33, then

$$5/(1 - 0.33) = 5/0.67 = 7.5 \text{ ml whole blood added}$$
$$\text{to 1 ml anticoagulant}$$

2. Mix gently by inversion and centrifuge at 135g for 15 min at room temperature to obtain platelet-rich plasma (PRP).

3. Draw off PRP using a plastic pasteur pipette and place in a clean plastic centrifuge tube.

4. Centrifuge residual blood at 1500g for 15 min to obtain autologous platelet-poor plasma (PPP).

5. Count PRP and dilute to 2.5×10^8/ml using autologous PPP.

Platelet-Rich Plasma Correction Formula

$$\frac{\text{Desired PRP count}}{\text{actual PRP count}} = \text{ml fraction of PRP in final 1 ml volume}$$

Bring the volume to 1.0 ml with PPP. Example:

$$\text{Desired PRP count} = 250,000/\text{mm}^3 \ (2.5 \times 10^8/\text{ml})$$
$$\text{Actual PRP count} = 450,000/\text{mm}^3$$

$$250,000/450,000 = 0.55 \text{ ml PRP}$$

$$0.55 \text{ ml PRP} + 0.45 \text{ ml PPP} = 1.0 \text{ ml PRP at } 250,000/\text{mm}^3$$

or

$$550 \text{ µl PRP} + 450 \text{ µl PPP} = 1.0 \text{ ml final volume PRP}$$

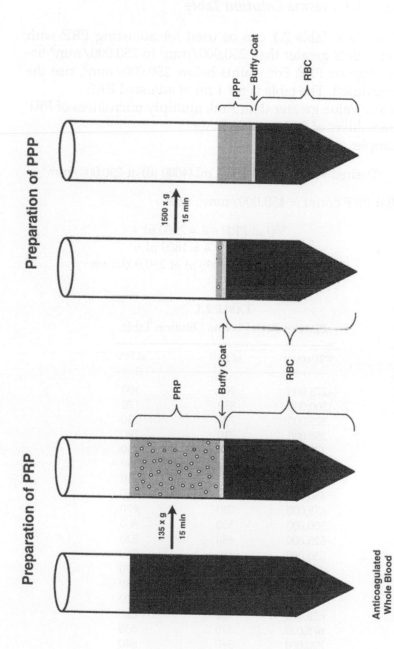

Figure 2.9. Preparation of PRP and PPP from a whole blood specimen.

Platelet-Rich Plasma Dilution Table

The following Table 2.1 can be used for adjusting PRP with platelet counts greater than $250,000/mm^3$ to $250,000/mm^3$ using autologous PPP. For counts below $250,000/mm^3$, use the PRP undiluted. The table is for 1 ml of adjusted PRP.

For any value greater than 1 ml, multiply microliters of PRP and microliters of PPP by that factor.

Example:

Desired volume of PRP = 4 ml (4000 µl) at $250,000/mm^3$

If initial PRP count = $450,000/mm^3$, use

$$550 \text{ µl PRP} \times 4 = 2200 \text{ µl} +$$
$$450 \text{ µl PPP} \times 4 = 1800 \text{ µl} =$$
a total volume of 4000 µl at $250,000/mm^3$

TABLE 2.1
Platelet-Rich Plasma Dilution Table

PRP count	µl PRP	µl PPP
275,000	900	100
300,000	830	170
325,000	770	230
350,000	710	290
375,000	670	330
400,000	620	380
425,000	590	410
450,000	550	450
475,000	530	470
500,000	500	500
525,000	480	520
550,000	450	560
575,000	430	570
600,000	420	580
625,000	400	600
650,000	380	620
675,000	370	630
700,000	360	640

PLATELET AGGREGATION AND SECRETION

In the laboratory, platelet function is most commonly assessed with an aggregometer. This instrument simply measures the increase in light transmission through a stirring suspension of platelets upon addition of exogenous agonist (Figure 2.10). Many instruments also calculate the rate at which aggregation occurs. Classically, it was thought that, without release of granular contents from the platelets, aggregation would be abnormal, showing only the primary or first wave of aggregation. If this kind of pattern was seen, it was assumed that either SPD or a release defect was the cause. Actual measurement of platelet ATP/ADP content and release were difficult and usually not done, and therefore a definitive diagnosis was not made. With the advent of lumiaggregometry, platelet release

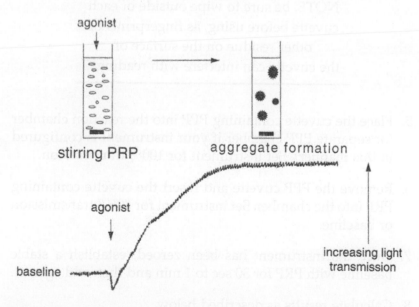

Figure 2.10. Aggregate formation and corresponding increase in light transmission as reflected in the aggregation tracing.

can be measured simultaneously with aggregation, and, by using a variety of potent agonists, the differentiation of SPD from release defects can be accomplished.

Platelet Aggregation Method

1. Draw and process blood as previously described.

2. Pipette 0.45 ml diluted PRP into an aggregation cuvette. This will be used to set "0%" transmission.

3. Pipette 0.45 ml PPP into a separate cuvette. This will be used to set "100%" transmission.

4. Add a stir bar to the above cuvettes and prewarm for 2 min.

NOTE: be sure to wipe outside of each cuvette before using, as fingerprints or other residue on the surface of the cuvette can interfere with readings.

---

5. Place the cuvette containing PPP into the reaction chamber (or separate PPP chamber if your instrument is configured in this manner). Set instrument for 100% transmission.

6. Remove the PPP cuvette and insert the cuvette containing PRP into the chamber. Set instrument for "0%" transmission or baseline.

7. Once the instrument has been zeroed, establish a stable baseline with PRP for 30 sec to 1 min and then add agonist.

8. Calculate results as described below.

How to Calculate Aggregation: Percent Maximum
Amplitude vs. Slope

There are two measurements that can be made using the platelet aggregometer. Traditionally, percent aggregation has been reported when discussing aggregation responses. Slope has been used in assays of von Willebrand factor ristocetin cofactor activity (vWF:RCo) and occasionally has been used when interpreting aggregation tracings. Both of these measurements are easily made manually, and many instruments now calculate these values automatically. Theoretically, aggregations should be very reproducible from laboratory to laboratory if they are done properly. Variability can be introduced when using an automated instrument if the instrument calculates aggregation at a preset time instead of at the maximal amplitude. In this case, if disaggregation occurs, aggregation estimates will be falsely low. To calculate maximum percent aggregation:

1. Measure the number divisions between the baseline and 100% on the chart paper (B).

2. Measure the number of divisions between baseline and maximum amplitude (A).

3. Divide A by B (A/B). This is percent maximal aggregation.

Figure 2.11 demonstrates the method for calculation of percent maximal aggregation.

In addition to quantification of percent maximal aggregation, it is useful to evaluate the pattern of aggregation tracings. PRP aggregations exhibit certain characteristics that may vary from agonist to agonist but are indicative of what is happening in the platelet as aggregation occurs. A representation of a "typical" aggregation tracing is depicted in Figure 2.12.

While aggregation calculations are the same from instrument to instrument industry-wide, slope calculations may

Calculation of Percent Maximal Aggregation

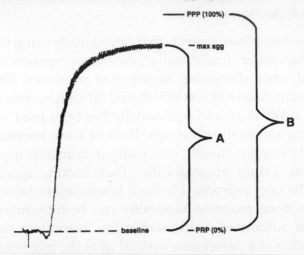

Figure 2.11. Calculation of percent maximal aggregation: A = number of chart divisions from baseline to maximal amplitude of aggregation tracing; B = number of chart divisions from baseline to theoretical "100%" aggregation; % aggregation = A/B.

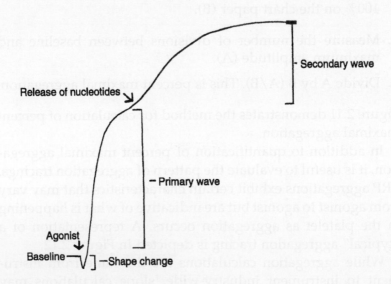

Figure 2.12. Description of events in "classic" biphasic aggregation.

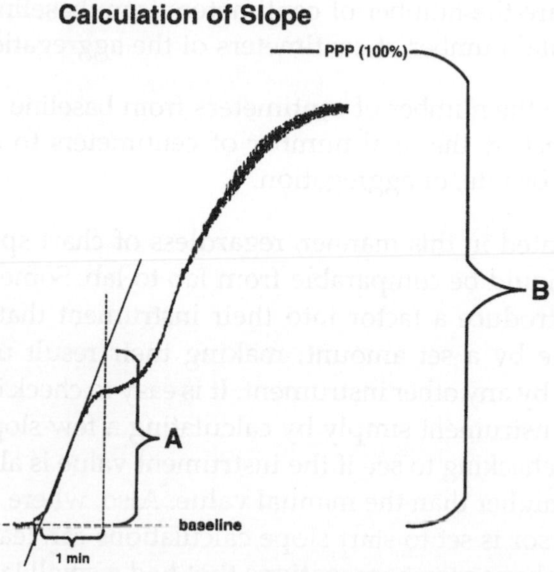

Calculation of Slope

Figure 2.13. Calculation of rate of aggregation or slope (M). A = distance from baseline to horizontal intersect of tangent drawn to steepest part of aggregation curve; B = distance from baseline to maximal height of y-axis (100% aggregation); M = A/B.

vary. The correct method for the calculation of slope is shown in Figure 2.13.

To calculate slope manually:

1. Draw a line tangent to the aggregation curve.

2. Determine how many centimeters your chart records in one minute.

3. Measure in centimeters from the point where the tangent intersects the baseline to the distance equal to 1 min.

4. Draw a line perpendicular to the baseline (x-axis) from this "1 minute" point to the intersect point of the tangent.

5. Measure the number of centimeters covered from baseline to this intersect point.

6. Measure the number of centimeters from baseline to 100% (the total number of centimeters of the aggregation y-axis).

7. Divide the number of centimeters from baseline to tangent intersect by the total number of centimeters to obtain the slope, or rate, of aggregation.

Calculated in this manner, regardless of chart speed, slope values should be comparable from lab to lab. Some manufacturers introduce a factor into their instrument that increases this value by a set amount, making their result unlike that obtained by any other instrument. It is easy to check if you own such an instrument simply by calculating a few slopes manually and checking to see if the instrument value is always a set amount higher than the manual value. Also, where the instrument cursor is set to start slope calculations can lead to problems. We have seen aggregations that had a small lag phase or a dip before the onset of full aggregation where the slope has actually been reported as zero.

Quality Control (QC) Tips for Platelet Aggregation

There are no commercially available QC kits for the aggregometer and no proficiency tests specific for laboratories performing aggregation testing. We who perform aggregations are expected to just "go on faith" that the instrument is working and is "calibrated." There is a way to check an aggregometer to see if it has actually been set to "0% transmission" (PRP) and "100% transmission" (PPP). Set the instrument per the manufacturer's instructions using PPP and PRP. Once the instrument has been set, establish a baseline with PRP, then remove the PRP cuvette and insert the PPP cuvette. The instrument should read "100% transmission." If it does not, there is a problem. We have found in our studies that there are some instruments that, even when set to, or when reading against, a PPP standard, will not read PPP as "100%." These instrument need to be serviced.

To take this test a step further, a "50% transmission" tube can be made by mixing PPP and PRP 1:1. After ascertaining that the aggregometer actually reads PPP as "100%," insert the "50%" tube; the tracing should deflect to halfway between the 0 and 100% readings. If your instrument is one that aborts the run if the tube is removed, set for "0" and "100" using 250 μl of both PPP and PRP and microvolume settings. Once the instrument is set, establish a PRP baseline and then gradually add 250 μl of PPP to the tube. The transmission reading should report approximately 50%.

Results of greater than 100% aggregation. In a properly QC'd and operating instrument, aggregations of more than "100%" should not be obtained. If a result greater than "100%" is measured, check the PPP for platelets. Aggregations of greater than "100%" are usually due to "laboratory error" in that the PPP actually contains enough residual platelets for the instrument to measure (Figure 2.14). PPP should have a platelet count of less than $5000/mm^3$.

Partial aggregation defects (or corrections). Occasionally the laboratory will obtain aggregation results that are not clearly indicative of any classic defect but which resemble partial defects. Usually in these cases, the collagen response will be partially decreased, ADP aggregation will not show a second wave, and epinephrine response will be minimal. In this situation, it should be suspected that the patient has ingested an aspirin or aspirin-containing medication within the past week to 10 days and has just forgotten about doing so. It is often difficult to obtain an accurate drug history from patients coming to the laboratory unless you have had the luxury of being able to schedule the platelet aggregation workup 2 weeks in advance and have advised the patient to abstain from all anti-platelet drugs. These post-aspirin responses are fairly predictable, and, if such responses are obtained, the test must be repeated. Whenever making a diagnosis of a platelet defect, it is imperative to confirm the diagnosis with a follow-up assay

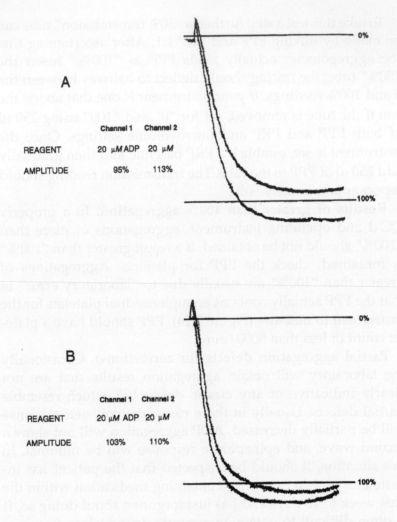

Figure 2.14. Two cases of aggregation tracings exhibiting greater than 100% aggregation. There are typically two situations that can produce results of greater than 100% aggregation: improper setting or calibration of the instrument, and residual platelets in the platelet-poor plasma used to set 100% aggregation. Since these two sets of tracings are each supposed to represent duplicate tests, "A" is probably due to improper calibration of channel 2 in this instrument; "B" illustrates what is seen when the PPP used to set an instrument contains residual platelets.

before labeling a patient as having an intrinsic platelet disorder.

Platelet Secretion

Platelet secretion studies take the investigation of platelet function one step further than simple aggregation testing. This tool can be very useful in the diagnosis of bleeding disorders such as SPD and release defect, especially in cases where the patients have the clinical bleeding in the absence of abnormal aggregation tracings usually associated with these disorders. Secretion studies can also be useful for the research laboratory that is investigating platelet activation or inhibition. One of the easiest and fastest ways to assess secretion is via the "lumiaggregometer," which measures platelet release simultaneously with platelet aggregation. When performing platelet release studies, it is imperative to construct a standard curve using known amounts of ATP prior to assessing the patient's release.

Lumiaggregation: Method

The basic setup for lumiaggregometry is the same as for aggregation, except that in every reaction tube a source of firefly luciferase (FFE) is added so that release of ATP can be measured simultaneously with aggregation. The most important aspect of performing release studies on the lumiaggregometer is establishment of a standard curve. Patient release values will be read directly from this curve. The standard curve is established using patient PRP and known amounts of ATP. To run the curve, set up the instrument as for an aggregation. Then:

1. Add FFE to the reaction cuvette.

2. Establish a baseline reading and then add 1 nM ATP to the cuvette. Repeat this step using 0.5 and 0.25 nM ATP. Each concentration should be run in duplicate.

3. Measure the number of chart units that the lumi-pen deflects (pen deflections) for each concentration, average the values, and plot on standard arithmetic graph paper or use a computer graphics program with the x-axis as nM ATP and the y-axis as chart units.

4. Add PRP and FFE to a reaction cuvette, establish a baseline, and add agonist.

5. Measure chart units on the lumi-channel.

6. Read from the standard curve.

An example of establishing a standard curve performed on a Payton lumiaggregometer is shown in Figure 2.15. The standard curve constructed from this method is also shown.

Other Methods Used for Measuring Granule Secretion

Platelet Dense Granule Secretion

^{14}C *serotonin release*

Platelets in plasma are loaded with ^{14}C serotonin (2 mCi/ml) at 37°C for 1 hr. Platelets are then gel-filtered or washed and the count adjusted to 2.5×10^8/ml with test buffer. Usually 1 μM imipramine is added to inhibit serotonin reuptake. An aliquot of platelet suspension, which will represent 100% total serotonin, is set aside for scintillation counting. After a test has been carried out, such as addition of agonist, a final of 1% formalin and 5 mM EDTA are added, the suspension is centrifuged at 15,000g for 10 min, and the supernatant is measured for ^{14}C serotonin. The percentage of release is determined by the ratio of the measured released serotonin to the total serotonin from an equivalent volume of platelets ×100.

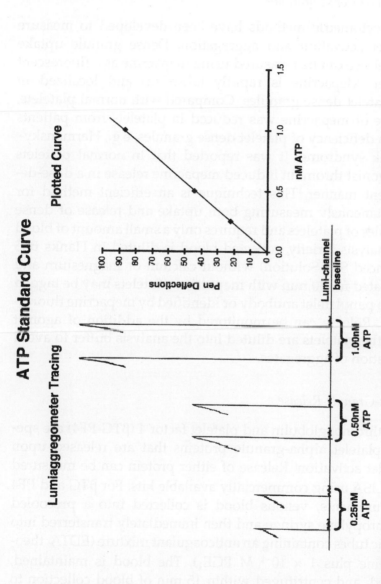

Figure 2.15. Examples of lumi-channel chart responses when known amounts of ATP are added to cuvettes containing PRP and FFE. Also shown is a standard curve constructed from these responses.

Flow cytometric assay for analysis of uptake and release
of platelet dense granules

Flow cytometric methods have been developed to measure platelet activation and aggregation. Dense granule uptake and release can be measured using mepacrine as a fluorescent marker. Mepacrine is rapidly taken up and localized in the platelet dense granules. Compared with normal platelets, uptake of mepacrine was reduced in platelets from patients with a deficiency of platelet dense granules, e.g., Hermansky–Pudlak syndrome. It was reported that in normal platelets the agonist thrombin induced mepacrine release in a dose-dependent manner. This technique is an efficient method for simultaneously measuring both uptake and release of dense granules of platelets and requires only a small amount of blood for analysis. Briefly, collected blood is diluted in Hanks BSS (Balanced Salt Solution) without calcium or magnesium and incubated for 30 min with mepacrine. Platelets may be tagged with a panplatelet antibody or identified by mepacrine fluorescence. Release can be monitored by the addition of agonist once the platelets are diluted into the analysis buffer to avoid formation of aggregates.

Alpha-Granule Release

Beta-thromboglobulin and platelet factor 4 (βTG-PF4) are specific platelet alpha-granule proteins that are released upon platelet activation. Release of either protein can be measured by ELISA using commercially available kits. For βTG and PF4 measurements, venous blood is collected into a precooled polypropylene syringe and then immediately transferred into plastic tubes containing an anticoagulant mixture (EDTA, theophylline plus 1×10^{-4} M PGE$_1$). The blood is maintained, chilled, and centrifuged within 15 min of blood collection to obtain the PPP for analysis. The PPP is carefully removed and stored at $-20°C$.

PLATELET AGONISTS

ADP. ADP concentrations in the range of 1 μM to 10 μM are usually used in assessment of platelet aggregation. At lower ADP concentrations (1–3 μM), either a single (monophasic) curve or a clearly biphasic curve thought to be due to the release of endogenous ADP will be seen. At the lowest concentrations, fibrinogen binding is usually not irreversible and disaggregation will occur. At higher ADP concentrations the biphasic response can be masked by the response elicited by the addition of endogenous agonist, but this is still considered a biphasic response because release has occurred. ADP aggregation with moderate concentrations of agonists is inhibited by ASA and other anti-platelet drugs due to inhibition of the cyclooxygenase pathway, and thus the inhibition of release of granular constituents.

BEWARE: If making up stock solutions
from bulk ADP, be sure to use the
molecular weight of the ADP received,
not the molecular weight of the free acid.

Epinephrine. Concentrations from 5 to 10 μM are typically used. With epinephrine, a small initial response followed by a larger full-scale response secondary to nucleotide release should be seen. As with ADP, this second wave of response is inhibited by ASA, NSAIDs, antihistamines, some antibiotics, and many other prescription and over-the-counter drugs. Epinephrine is the least consistent agonist used in the assessment of platelet aggregation, and, if the epinephrine response is the only abnormality seen in testing, one should be very hesitant to make the diagnosis of a "disorder" based on this result.

Collagen. Several forms of collagen are available commercially. Equine collagen is typically used at concentrations ranging from 1–5 μg/ml. Collagen is the strongest of the typical agonists used in the clinical lab. A typical lag phase of approximately 1 minute is seen during which the platelets are adhering to the collagen fibrils and undergoing shape change, and release. The aggregation response seen is really the "second wave" of aggregation subsequent to the release reaction. Thus, at low concentrations of collagen, ASA and other anti-platelet drugs will totally inhibit the aggregation response.

Arachidonic Acid (AA). In the presence of cyclooxygenase, AA is converted to thromboxane A_2, which is a potent platelet agonist. Aspirin inhibits the cyclooxygenase pathway and therefore inhibits aggregation in response to arachidonic acid. Patients who have ingested aspirin or other anti-platelet drugs or who have an intrinsic release defect or Glanzmann's thrombasthenia will have abnormal AA aggregation. Those patients with SPD should exhibit a normal response to AA.

Ristocetin. This antibiotic causes platelet aggregation in the presence of normal platelets and a normal complement of vWF antigen. Abnormal aggregation in response to ristocetin could be indicative of Bernard–Soulier Syndrome (absence of GPIb) or vWD, and deserves further evaluation. Abnormal ristocetin aggregation has also been reported in SPD. Ristocetin is usually used at a concentration of 1.5 mg/ml in PRP studies.

α-Thrombin. This is a very potent agonist with one major complication in PRP or whole blood aggregation testing: cleavage of fibrinogen and formation of a clot. γ-Thrombin can be used for aggregation in place of α-thrombin, but it is not readily available commercially. If used in platelet aggregation or activation studies, the concentration of γ-thrombin used should be approximately 0.024 units/ml. α-Thrombin may be used to activate platelets in washed or gel-filtered platelet preparations. Usually 0.1–0.5 units/ml of α-thrombin is used for these studies.

TRAP. Thrombin receptor agonist peptide (TRAP) is a small peptide that corresponds to the amino acid sequence of the "tethered ligand" that is generated after thrombin hydrolysis of the N-terminus region of the thrombin receptor. The addition of this peptide sequence to platelets mimics very strong activation seen with thrombin without the complication of fibrinogen cleavage and clot formation. TRAP will give normal responses in most platelet defects except for Glanzmann's thrombasthenia. It is now being used in platelet aggregation testing to monitor the pharmacodynamic effects of new anti-platelet drugs that block fibrinogen binding to the platelet. Typical concentrations used for aggregation studies are 5 or 10 μM.

mAb7. This anti-CD9 monoclonal antibody activates platelets via crosslinking of CD9 and the platelet FCγRII receptor. It is very useful in differentiating SPD (abnormal release values) from release-type defects (normal release values). Patients with Glanzmann's thrombasthenia will have an abnormal aggregation response to mAb7 but will exhibit normal release.

D3. D3, an anti-GPIIIa monoclonal antibody, activates the GPIIb–IIIa complex to bind fibrinogen but does not activate the platelet. When added to a stirring suspension of platelets, this antibody induces a small wave of platelet aggregation (25–35%). In addition, GPIIb–IIIa conformational states and post-receptor occupancy events have been examined using the unique binding properties of this mAb.

In summary, as with any test used in the clinical lab, a normal range should be established for aggregation and release with each agonist at each concentration. Table 2.2 is provided as a *guideline*, but the values may vary depending on the laboratory instrumentation, normal population, specimen handling, etc.

CLOT RETRACTION

Clot retraction may be measured in either whole blood or PRP. It is dependent on the interaction of platelets with fibrinogen.

TABLE 2.2
Table of Normal Ranges

Agonist	Percent aggregation	nmoles released ATP
10 μM ADP	71–88	0.40–0.64
5 μM ADP	69–88	0.41–0.63
1 μM ADP	13–58	0.00–0.47
0.5 μM ADP	7–23	0.00–0.00
2 μg/ml collagen	70–94	0.46–0.74
1 μg/ml	26–98	0.26–0.72
5 μM epinephrine	78–88	0.40–0.52
0.5 μM	17–91	0.00–0.64
0.024 units/ml γ-thrombin	65–96	0.80–1.6
20 μg/ml mAb7	63–96	0.78–1.56
5 μM TRAP	63–96	0.78–1.56

An abnormal clot retraction may reflect abnormal platelet numbers, abnormal platelet glycoprotein amount or structure, abnormal platelet signaling mechanisms, abnormal fibrinogen levels, or abnormal fibrinogen structure. The assessment of clot retraction inhibition may be a useful tool when looking at new anti-platelet drugs, platelet antibodies, or unexplained bleeding problems.

Whole Blood Clot Retraction

1. Place 5 ml whole blood in a clean graduated glass conical centrifuge tube (at this time a microhematocrit must be performed).

2. Cap tube with parafilm and put wooden applicator stick through the top, down into blood.

3. Place tubes at 37°C for 24 hr.

4. Remove tubes from bath and carefully remove clot from tube.
5. Centrifuge the tube at 1500*g* for 5 min to pack residual RBC.
6. Record the following data:

   a. Total volume of blood added to tube.

   b. Residual volume after clot removal.

   c. Volume of packed RBC.

   d. Serum volume (b – c = d).

   e. Hematocrit

$$\frac{\text{Total volume} - \text{serum volume}}{\text{Total volume}} \times 100 - \text{Hct} = \% \text{ Fluid volume of the clot}$$

Normal %FVC ≤ 20%. (An example of whole blood clot retraction is shown in Figure 2.16; see color plate.)

PRP Clot Retraction

1. To 10 × 75 glass tubes, add 0.5 ml PRP + 0.05 ml 0.25 *M* $CaCl_2$.

   If testing for clot retraction inhibition, add 0.05 ml substance being tested before addition of PRP or $CaCl_2$.
2. Mix and incubate at 37°C for 1 hr.
3. Observe clot every 15 min until retraction is complete. Make note of percentage clot retraction in comparison to untreated or normal tube.

We have devised a grading system for PRP clot retraction to help in the "quantification" of results.

++++	90–100% clot retraction
+++	60–90% clot retraction
++	30–60% clot retraction
+	less than 30% clot retraction
—	total inhibition of clot retraction.

Figure 2.17 represents a typical PRP clot retraction assay.

Figure 2.17. PRP clot retraction. Tubes are arranged from complete clot retraction (far left) to total inhibition of clot retraction (far right).

PLATELET ADHESION

Platelet adhesion to immobilized ligand is generally carried out on polystyrene plates or on microscope slides precoated for 16 hr with ligand (generally 10–100 µg/ml) or with BSA (5 mg/ml) in PBS. The plates are then washed three times with PBS, blocked for 2 hr with BSA to block ligand-free sites, and washed again with PBS. Aliquots of platelets are then added to the plates typically for 30–60 min at room temperature. If collagen is used as the ligand, it is recommended that acid-soluble type I calf skin collagen be dissolved in 0.5% acetic acid and only applied to plates for 2 hr. Platelet adhesion to collagen must be carried out in the presence of 2 mM MgCl$_2$, whereas adhesion to other ligands such as fibrinogen is usually performed in the presence of millimolar calcium.

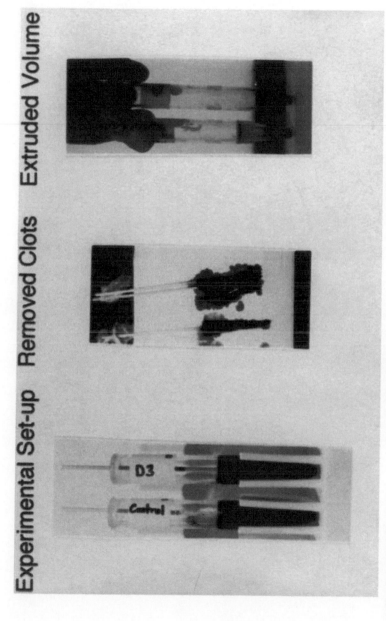

Figure 2.16. Whole blood clot retraction. The control tube exhibits normal clot retraction; the tube treated with D3 demonstrates clot retraction inhibition.

FLOW CYTOMETRIC ANALYSIS OF PLATELET SURFACE PROTEINS

Labeling platelets with antibodies directed against surface antigens and analyzing antibody binding by flow cytometry is a rapid and sensitive technique for both research and clinical laboratory studies. Flow cytometry may be used to evaluate the activation state of the platelet, for measuring platelet interactions with other cell types, for assessment of microparticle formation, and for antigen surface density. Figure 2.18 is a flow cytometric analysis of GPIb and GPIIb–IIIa surface density on normal platelets (MMW) and platelets from a Glanzmann's thrombasthenia (GT) individual (ETT) who has about 20–25% of normal levels of GPIIb–IIIa. The residual GPIIb–IIIa cannot function to mediate platelet aggregation. The figure demonstrates a deficiency of GPIIb–IIIa on the GT platelets but comparable levels of surface GPIb (Table 2.3). The critical factors in this type of surface antigen analysis are obtaining the appropriate antibodies for the test and optimizing binding of these reagents. Antibody binding may be carried out using whole

TABLE 2.3

Mean Fluorescence Intensity of Antibody
Staining to Normal Platelets and Platelets from
a Glanzmann's Thrombasthenic Patient

Antibody	GT	Normal
Mouse IgG	3.0	3.7
D3 (anti-GPIIIa LIBSs)	15.6	44.3*
AP3 (ant-GPIIIa)	17.8	135.2
7E3(anti-GPIIb–IIIa)	19.9	135.0
6D1(anti-GPIb)	104.1	108.7

*D3 binding is equivalent to that observed with AP3 and 7E3 only when ligand is bound to GPIIb–IIIa.

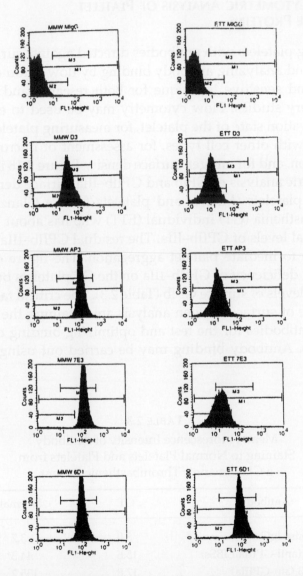

Figure 2.18. Flow cytometric analysis of surface density of GPIb and GPIIb–IIIa on normal (MMW) and Glanzmann's thrombasthenic (ETT) platelets. MIgG, a mouse IgG control; D3 and AP3, anti-GPIIIa antibodies (AP3 was kindly provided by Dr. Peter Newman); 7E3 (anti-GPIIb–IIIa) and 6D1 (anti-GPIb antibody) generously donated by Dr. Barry Coller.

blood, PRP, or washed platelets. Whole blood and PRP require very little sample volume. If whole blood is used, the platelets must be tagged by a panplatelet antibody to discriminate the platelet population from the other blood cells. Procedures for flow cytometric analysis of platelet surface antigens are outlined in the Appendix.

RECEPTOR OCCUPANCY

Historically, patients with "hyperaggregable platelets" were treated with such anti-platelet drugs as ASA or NSAIDs and were not monitored by the laboratory to assess inhibition of the platelet response. As understanding of platelet function and physiology deepened, identification of the RGD binding site of fibrinogen on platelet GPIIb–IIIa was made and the focus shifted from general platelet inhibition to the specific blockade of fibrinogen binding. The introduction of peptidomimetics and their ability to totally block all platelet aggregation (instead of just the "second wave") led to a need for a method of monitoring patient response to these drugs. Platelet aggregation can be and still is used, but a more sensitive and direct measure of fibrinogen receptor blockade was needed. Since there is no currently available way to directly measure the presence of these small peptides on the platelet, an indirect approach was developed that takes advantage of the ability of many of these drugs to induce ligand-induced binding sites or LIBSs.

LIBSs can be detected using anti-LIBS antibodies and a flow cytometer. The more "ligand" or antagonist bound to the GPIIb–IIIa of the platelet, the more antibody will bind until a saturation point is reached. Using this knowledge, a receptor occupancy (RO) assay was developed to measure the percentage of patient platelet GPIIb–IIIa occupied by "ligand" (or drug), which is reflected in the binding of a phycoerythrin-conjugated anti-LIBSs, D3. This binding is compared to a sample to which saturating amounts of drug are added *in vitro* to measure total surface GPIIb–IIIa (Figure 2.19). Of course, for each new drug, dose–response studies have to be performed to establish that increasing amounts of drug added *in vitro* lead to increasing amounts of bound antibody until saturation. These studies are compared to inhibition of platelet aggregation. One key feature of the RO assay is that it actually measures blocked receptors and the result is not influenced by the presence of any additional anti-platelet drugs. These drugs can affect the results of inhibition of aggregation tests because baseline values can be depressed due to their presence. Further discussion of GPIIb–IIIa antagonists and receptor occupancy is found in Chapter 4.

Receptor Occupancy Measurement

Whole blood or PRP samples can be used for receptor occupancy assays. If a whole blood sample is used, a dual-labeling technique must be used to distinguish platelets from the other blood cells. This panplatelet tag may be another GPIIb–IIIa antibody or an antibody that binds a different platelet-specific antigen. Regardless of the target antigen, this antibody binding must not be interfered with by the presence of drug or binding of the second antibody that will report receptor occupancy.

For PRP, the sample is divided into aliquots. The assay includes binding of a direct fluorochrome conjugate (IgG) as a negative control and of a direct conjugate (D3) to measure receptors with bound drug. D3 binding is also measured in a PRP aliquot with added excess drug that represents total GPIIb–IIIa receptors. The mean fluorescence intensity is

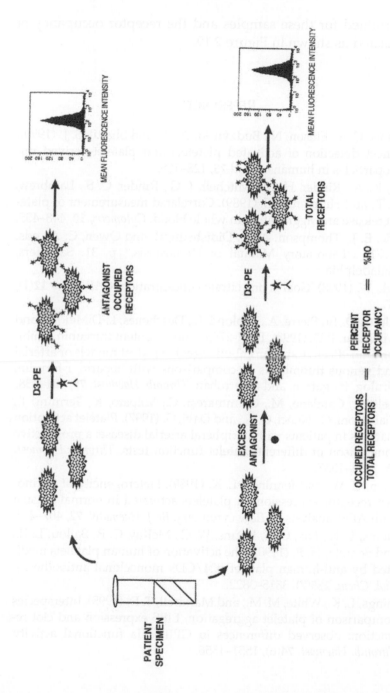

Figure 2.19. Schematic representation of receptor occupancy binding assay using D3 and flow cytometry.

determined for these samples and the receptor occupancy is calculated as shown in Figure 2.19.

REFERENCES

Abrams, C. S., Ellison, N., Budzynski, A. Z., and Shattil, S. J. (1990). Direct detection of activated platelets and platelet-derived microparticles in humans. *Blood* **75**, 128–138.

Ault, K. A., Rinder, H. M., Mitchell, J. G., Rinder, C. S., Lambrew, C. T., and Hillman, R. S. (1989). Correlated measurement of platelet release and aggregation in whole blood. *Cytometry* **10**, 448–455.

Bowie, E. J., Thompson, A. R., Didisheim, P., and Owen, C. A., eds. (1971). "Laboratory Manual of Hemostasis," p. 31. Saunders, Philadelphia.

Check, W. (1998). Goal is one citrate concentration. *CAP Today* **12**(1), 27.

Finkle, C. D., St. Pierre, A., Leblond, L., Deschenes, I., DiMaio, J., and Winocour, P. D. (1998). BCH-2763, a novel potent thrombin inhibitor, is an effective antithrombotic agent in rodent models or arterial and venous thrombosis — comparisons with heparin, r-hirudin, hirulog, inogatran, and argatroban. *Thromb. Haemost.* **79**, 431–438.

Gresele, P., Catalano, M., Giammaresi, C., Volpato, R., Termini, T., Ciabattoni, G., Nenci, G. G., and Davi, G. (1997). Platelet activation markers in patients with peripheral arterial disease: a prospective comparison of different platelet function tests. *Thromb. Haemost.* **78**, 1434–1337.

Jackson, C. W., and Jennings, L. K. (1989). Heterogeneity of fibrinogen receptor expression on platelets activated in normal plasma with ADP: analysis by flow cytometry. *Br. J. Haematol.* **72**, 407–414.

Jennings, L. K., Fox, C. F., Kouns, W. C., McKay, C. P., Ballou, L. R., and Schultz, H. E. (1990). The activation of human platelets mediated by anti-human platelet p24/CD9 monoclonal antibodies. *J. Biol. Chem.* **265**(7), 3815–3822.

Jennings, L. K., White, M. M., and Mandrell, T. D. (1995). Interspecies comparison of platelet aggregation, LIBS expression and clot retraction: observed differences in GPIIb–IIIa functional activity. *Thromb. Haemost.* **74**(6), 1551–1556.

Kouns, W. C., Wall, C. D., White, M. M., Fox, C. F., and Jennings, L. K. (1990a). A conformation-dependent epitope of glycoprotein IIIa. *J. Biol. Chem.* **265**(33), 20594–20601.

Kouns, W. C., Rubenstein, E., Carroll, R. C., White, M. M., and Jennings, L. K. (1990b). Induction of antibody–FCγRII-mediated activation through alteration of the cytoskeleton. *Circulation* **82**(4), 111.

Mondoro, T. H., Wall, C. D., White, M. M., and Jennings, L. K. (1996). Selective induction of a glycoprotein IIIa ligand-induced binding site by fibrinogen and von Willebrand factor. *Blood* **88**(10), 3824–3830.

Pignatelli, P., Pulcinelli, F. M., Ciatti, F., Pesciotti, M., Sebastiani, S., Ferroni, P., and Gassaniga, P. P. (1995). Acid citrate dextrose (ACD) formula A as a new anticoagulant in the measurement of *in vitro* platelet aggregation. *J. Clin. Lab. Anal.* **9**(2), 138–140.

Ts'ao, C. H., Lo, R., and Raymond, J. (1976). Critical importance of citrate–blood ratio in platelet aggregation studies. *Am. J. Clin. Pathol.* **65**(4), 518–522.

Vostal, J. G. and Mondoro, T. H. (1997). Liquid cold storage of platelets: a revitalized possible alternative for limiting bacterial contamination of platelet products. *Transfus. Med. Rev.* **11**, 286–295

Wall, J. E., Buijs-Wilts, M., Arnold, J. T., Wang, W., White, M. M., Jennings, L. K., and Jackson, C. W. (1995). A flow cytometric assay using mepacrine for the study of uptake and release of platelet dense granule contents. *Br. J. Haem.* **89**, 380.

White, M. M., Foust, J. T., Mauer, A. M., Robertson, J. T., and Jennings, L. K. (1992). Assessment of lumiaggregometry for research and clinical laboratories. *Thromb. Haemost.* **76**(5), 572–577.

FUNCTIONAL
ABNORMALITIES
OF PLATELETS

Introduction

**Storage Pool Disorders (SPDs) and Release
Defects**

> Storage Pool Disorders
> Release Defects

INTRODUCTION

There are many excellent in-depth reviews of platelet function abnormalities. The purpose of this book is not to duplicate these works but rather to provide investigators in the field of platelet research and clinical laboratory personnel given the job of assessing platelet function with a resource for working with platelets in the laboratory.

Briefly and simply, platelet function abnormalities can be grouped under two headings:

1. Diminished platelet responses to weak agonists (storage pool defect and release defect or aspirin-like syndromes).

2. Absent platelet aggregation to all agonists (Glanzmann's type defects — abnormalities or absence of GPIIb–IIIa). This type of response may also be seen in patients with abnormal or absent fibrinogen.

STORAGE POOL DISORDERS (SPDs) AND RELEASE DEFECTS

Classically, storage pool disorders and release defects are associated with an abnormal second wave of aggregation secondary to abnormal release of nucleotides from the storage granules. Reports have been made of several hundred cases of storage pool-like defects with abnormal bleeding but normal platelet aggregation. The only way of making the diagnosis on these patients was via adenine nucleotide studies. The etiology of the various storage pool syndromes and release defects are discussed at length in various texts. We include a list of the types of syndromes in which SPD and release defects have been identified.

Storage Pool Disorders

Dense granule storage pool disease γ-SPD:

Hermansky–Pudlak syndrome (oculocutaneous albinism)

Chediak–Higashi syndrome (partial oculocutaneous albinism)

Wiskott–Aldrich syndrome

TAR syndrome

Alpha/dense granule storage pool disease: αγ-SPD

Alpha-granule deficiency: α-SPD

Gray platelet syndrome

Release Defects

Aspirin-like syndromes

Cyclooxygenase deficiency

Thromboxane A_2 receptor defects

Drug-induced

Aspirin (ASA)

Nonsteroidal antiinflammatory drugs (NSAIDs)

Antibiotics

Dipyridamole

For an excellent review of all drugs affecting platelet function, refer to "Hematology: Basic Principles and Practice" (Hoffman et al., 1995).

There are other disorders in which acquired platelet function defects have been reported. These defects range from storage pool and release defects to abnormalities in platelet adhesion. Some of the disorders that may manifest these defects are:

> Uremia
> Preleukemia and acute leukemias
> Myeloproliferative disorders
> Dysproteinemias
> Liver disease
> Anti-platelet antibodies

Glanzmann's Thrombasthenia (GT). Glanzmann's patients have a deficiency or mutation in the GPIIb–IIIa receptor. While absence of GPIIb–IIIa yields a definitive diagnosis of GT, variants have been described that demonstrate platelet GPIIb–IIIa but have a mutation that renders it nonfunctional. The diagnostic features are: absent platelet aggregation to ADP, collagen, epinephrine, and thrombin, and usually an abnormal clot retraction.

von Willebrand's Disease. Of course, no discussion of platelet abnormalities is complete without including von Willebrand's disease. Patients with von Willebrand's disease exhibit bleeding similar to that seen with intrinsic platelet defects. The aggregation responses in these individuals are normal to all agonists, except possibly ristocetin. The defect in this case is not in the platelet, but in the von Willebrand factor molecule that binds to GPIb–IX and supports platelet adhesion to the subendothelium. Diagnosis of this syndrome is made by assessing Factor VIII coagulant activity (FVIII:C), von Willebrand factor antigen activity (vWF:Ag), and vWF ristocetin cofactor activity (vWF:RCo). Assessment of the multimeric structure of the von Willebrand antigen is useful in determining the specific type of von Willebrand's disease that is present. A new method of assessing the Factor VIII binding activity of vWF has been reported that should also help delineate specific subtypes of vWD.

Bernard–Soulier Syndrome (BSS). BSS patients have a deficiency of GPIb–IX–V, so that their platelets cannot interact with vWF. This leads to a problem with platelet adhesion to the subendothelium. These individuals exhibit bleeding very similar to that seen in von Willebrand's disease, but they also usually exhibit large platelets and borderline thrombocytopenia upon examination of the peripheral smear. The aggregation responses seen in BSS are normal to all agonists but ristocetin. If using only platelet aggregation to make a diagnosis of a bleeding disorder, one could mistake BSS for von Willebrand's disease. To distinguish these disorders, Factor VIII coagulant activity (FVIII:C), von Willebrand factor antigen activity (vWF:Ag), and von Willebrand factor ristocetin cofactor activity (vWF:RCo) — all or any combination of which would be decreased in vWD — must be assessed. In addition, flow cytometric analysis of GPIb–IX–V surface density would aid in diagnosis.

The aggregation and release patterns classically seen in normal patients and in the patient types described above are shown in Figures 3.1–3.4.

Bleeding Time Prolongation. Platelet function disorders can cause prolongation of bleeding time; however, bleeding time measurement is not considered a good predictor of the probability of bleeding. The general recommendation is that only when a patient is suspected of having a bleeding disorder should a bleeding time measurement be considered. When a patient's platelet count is less than $100,000/mm^3$, the bleeding time measurement has sometimes been used as an assessment of platelet function since aggregometry values are not reliable at counts less than $100,000/mm^3$. Generally speaking, an inverse relationship appears to exist between the platelet count and the bleeding time when the counts range from $100,000/mm^3$ down to $10,000/mm^3$. Bleeding times of greater than 30 min are usually seen with platelet counts of less than $10,000/mm^3$. It is possible that a disproportionally long bleeding time in relationship to platelet count may suggest a platelet function defect, whereas a disportionate shortening of time may suggest increased platelet responsiveness. Many variables can affect the outcome of the bleeding time, including the experience of the laboratory staff, the venous pressure applied,

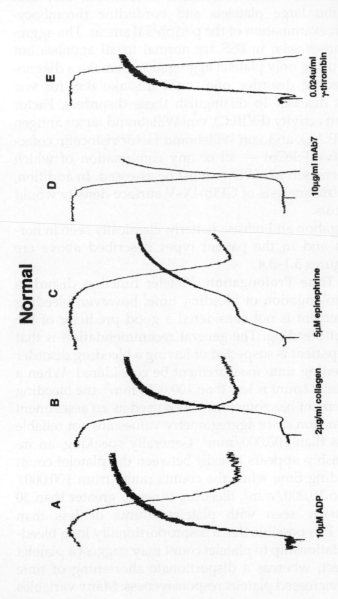

Figure 3.1. Aggregation and release tracings from a normal donor reflecting responses to typical agonists used in the laboratory. Deflection in baseline tracing indicates point at which agonist is added. Aggregation response is lower tracing; release response is upper tracing. Note classic biphasic response of epinephrine. ADP second wave is masked by strong response to exogenous agonist. Lumi-responses to both mAb7 and γ-thrombin are off the scale.

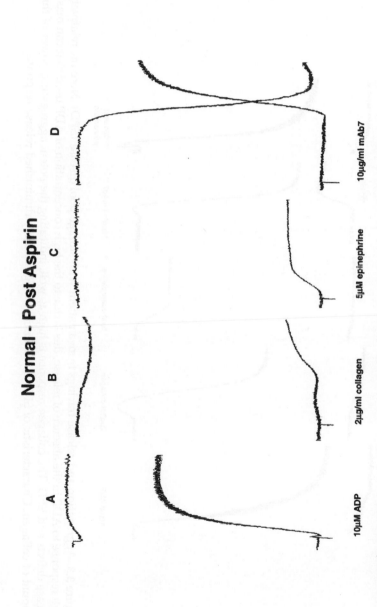

Figure 3.2. Aggregation and release tracings from a normal donor post-aspirin ingestion. Note absence of release and full-scale aggregation in agonists acting through cyclooxygenase pathway, and strong release in agonist that activates through FCγRII.

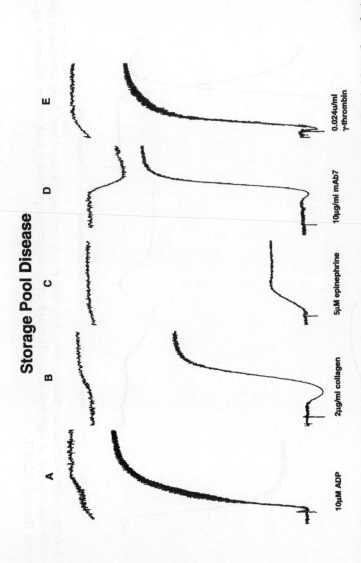

Figure 3.3. Aggregation and release tracings from a patient with storage pool disease (SPD). Note the relatively high response to ADP in the absence of release; this is due to the high concentration of ADP used that can mask release defects and SPD. The collagen response is high as well, indicating the potent aggregating power of this amount of collagen. Characteristic of SPD is the absent or significantly diminished release response.

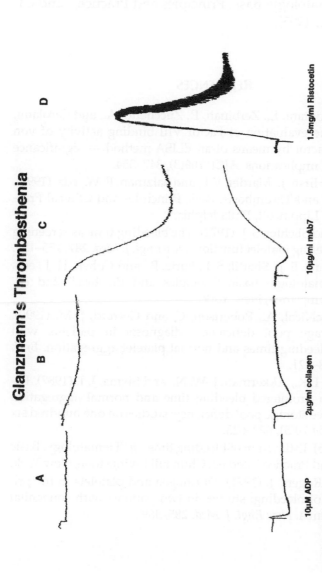

Figure 3.4. Aggregation and release tracings from patient with Glanzmann's thrombasthenia. Note in curve using mAb7 that the "aggregation" response seen is actually lysis of platelets rather than aggregation. Platelet lysis results in an increase in light transmission registered as "aggregation," but there is no real aggregation seen visually. The release effect seen here is also due to lysis, although in GT release does occur in response to strong agonists such as mAb7 and thrombin. The ristocetin aggregation response reflects the initial agglutination of platelets induced by vWF binding to GPIb but no aggregation is seen because fibrinogen cannot bind.

and the type of incision made during performance of the technique. A thorough discussion of bleeding time is provided in "Hematology: Basic Principles and Practice," 2nd ed. (Hoffman *et al.*, 1995).

REFERENCES

Casonato, A., Pontara, E., Zerbinati, P., Zucchetto, A., and Girolami, A. (1998). The evaluation of Factor VIII binding activity of von Willebrand factor by means of an ELISA method — significance and practical implications. *AJCP* **109**(3), 347–354.

Colman, R. W., Hirsh, J., Marder, V. J., and Salzman, E. W., eds. (1994). "Hemostasis and Thrombosis: Basic Principles and Clinical Practice," 3rd ed. Lippincott, Philadelphia.

Harker, L. A, and Slichter, S. J. (1972). The bleeding time as screening test for evaluating platelet function. *New Engl. J. Med.* **287**, 155–159.

Hoffman, R., Benz, E. J., Shattil, S. J., Furie, B., and Cohen, H. J., eds. (1995). "Hematology: Basic Principles and Practice," 2nd ed. Churchill Livingstone, New York.

Israels, S. J., McNicol, A., Robertson, C, and Gerrard, J. M. (1990). Platelet storage pool deficiency: diagnosis in patients with prolonged bleeding times and normal platelet aggregation. *Br. J. Haem.* **75**, 118–121.

Nieuwenhuis, H. K., Akkerman, J.-W. N., and Sixma, J. J. (1987). Patients with a prolonged bleeding time and normal aggregation tests may have storage pool deficiency: studies on one hundred six patients. *Blood* **70**(3), 620–623.

Sixma, J. J. (1995). Estimation of bleeding time. *In* "Hematology: Basic Principles and Practice," 2nd ed. Churchill Livingstone, New York.

Weiss, H., and Rogers, J. (1971). Fibrinogen and platelets in the primary arrest of bleeding: studies in two patients with congenital afibrinogenemia. *New Engl. J. Med.* **285**, 369.

THERAPEUTIC
APPROACHES TO
INHIBITION OF
PLATELET FUNCTION

INTRODUCTION

Platelets are a key component of arterial thrombi, and an effective pharmacological inhibition of platelet aggregation is a primary goal of antithrombotic therapy. In the last several years, effort has been devoted to generation of GPIIb–IIIa antagonists that can be utilized in the treatment of acute coronary syndromes. These antagonists are potentially more effective than aspirin or ticlopidine, as blockade of GPIIb–IIIa would be an effective inhibitor of platelet aggregation to all agonists rather than limited to only the weak agonists, as observed for ASA inhibition of platelet aggregation and to ADP for ticlopidine-mediated platelet aggregation inhibition.

GPIIb–IIIa ANTAGONISTS

GPIIb–IIIa antagonists bind the receptor with high affinity and inhibit the binding of fibrinogen and vWF upon platelet activation. Platelet aggregation is effectively blocked when the drug is bound to GPIIb–IIIa (Figure 4.1). An advantage of i.v. administration of a GPIIb–IIIa receptor blockade is its rapid binding to the receptor, with pharmacodynamic effects observed within minutes. With peptides and peptidomimetics, termination of the infusion results in progressive return of platelet function, should rescue of platelet function be desirable. Bioavailable GPIIb–IIIa inhibitors are also being evaluated in these coronary syndromes.

Pharmacodynamic evaluation of GPIIb–IIIa antagonists has depended mainly upon *ex vivo* measurement of inhibition of platelet aggregation. As already outlined herein, platelet aggregometry is beneficial for identifying qualitative platelet defects. This technique requires highly specialized personnel and an available aggregometry laboratory. Since aggregometry responses can have significant intrapatient and interpatient variability, an alternative method for assessment of receptor blockade has been developed. This method is called "receptor

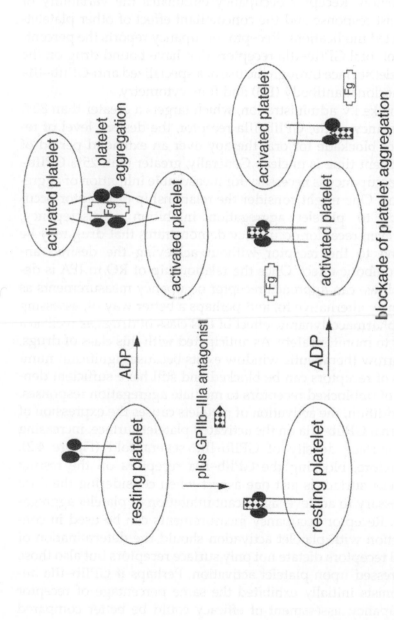

Figure 4.1. Blockade of fibrinogen binding and platelet aggregation by GPIIb–IIIa receptor antagonists.

occupancy" and is not synonymous with platelet aggregation inhibition. Receptor occupancy eliminates the variability of agonist response and the concomitant effect of other platelet-directed medications. Receptor occupancy reports the percentage of total GPIIb–IIIa receptors that have bound drug on the platelet surface through the use of a specialized anti-GPIIb–IIIa monoclonal antibody (D3) and flow cytometry.

Unlike i.v. administration, which targets a greater than 80% occupancy of the GPIIb–IIIa receptor, the desired level of receptor blockade for oral therapy over an extended period of treatment time is unclear. Generally, greater than 25% GPIIb–IIIa occupancy is necessary for measurable inhibition of aggregation. One might consider the relationship of receptor occupancy to platelet aggregation inhibition a discrepancy; however, receptor occupancy demonstrates that drug may be bound to the receptor without achieving the desired antithrombotic effect. Once the relationship of RO to IPA is defined, we can approach receptor occupancy measurements as the new alternative to, and perhaps a better way of, assessing the pharmacodynamic effect of this class of drugs as well as a way to monitor safety. As anticipated with this class of drugs, a narrow therapeutic window exists because significant numbers of receptors can be blocked and still have sufficient density of unblocked receptors to mediate aggregation responses. In addition, the activation of platelets causes the expression of internal GPIIb–IIIa on the activated platelet surface, increasing the surface density of GPIIb–IIIa several-fold (Figure 4.2). Therefore, blocking the GPIIb–IIIa receptors on the resting platelet surface is just one aspect when considering the dose necessary to achieve significant inhibition of platelet aggregation. Receptor occupancy measurements can be used in conjunction with platelet activation should the determination of total receptors dictate not only surface receptors but also those expressed upon platelet activation. Perhaps if GPIIb–IIIa antagonists initially exhibited the same percentage of receptor occupancy, assessment of efficacy could be better compared from antagonist to antagonist for specific coronary syndromes.

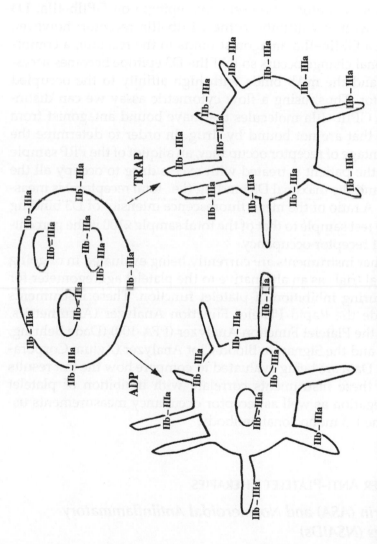

Figure 4.2. Increase in GPIIb–IIIa surface density after activation by either ADP or thrombin. The relative increase in surface expression is dependent upon the strength of agonist.

The receptor occupancy assay currently used in several on-going trials is the D3 monoclonal antibody binding assay. This assay utilizes the unique binding properties of the D3 mAb that binds to a conformation-sensitive epitope on GPIIb–IIIa. D3 binds with low affinity to the GPIIb–IIIa receptor; however, when a GPIIb–IIIa antagonist binds to the receptor, a conformational change occurs so that the D3 epitope becomes accessible and the mAb binds with high affinity to the occupied receptor. Thus, using a flow cytometric assay we can distinguish GPIIb–IIIa molecules that have bound antagonist from those that are not bound by drug. In order to determine the percentage of receptor occupancy, an aliquot of the PRP sample from the patient is treated with excess drug to occupy all the sites and the maximal D3 binding (i.e., total receptors) is measured. A ratio of the mean fluorescence intensity of D3 binding to the test sample to that of the total sample ×100 is the percentage of receptor occupancy.

Other instruments are currently being evaluated in ongoing clinical trials as an alternative to the platelet aggregometer for measuring inhibition of platelet function. These instruments include the Rapid Platelet Function Analyzer (Accumetrics, Inc.), the Platelet Function Analyzer (PFA-100) (Dade Behring, Inc.), and the Signature Blood Clot Analyzer (Xylum Corporation). Data are being evaluated to compare how the test results from these instruments correlate with inhibition of platelet aggregation as well as receptor occupancy measurements using the D3 monoclonal antibody.

OTHER ANTI-PLATELET THERAPIES

Aspirin (ASA) and Nonsteroidal Antiinflammatory Drugs (NSAIDs)

Aspirin, the primary anti-platelet drug prescribed, has been demonstrated to reduce the risk of arterial thrombosis compared to placebo. ASA and other nonsteroidal antiinflamma-

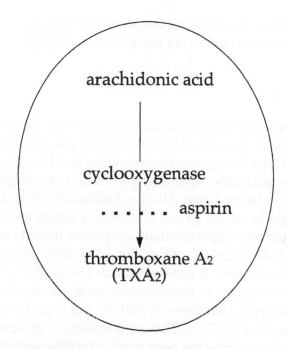

arachidonic acid

cyclooxygenase

. . . . . . aspirin

thromboxane A2
(TXA2)

Figure 4.3. Inhibition of thromboxane A_2 formation by aspirin. Aspirin irreversibly inhibits the cyclooxygenase enzyme, inhibiting conversion of arachidonic acid to thromboxane A_2.

tory medications inhibit the cyclooxygenase pathway. Only ASA irreversibly inhibits cyclooxygenase and prevents thromboxane A_2 synthesis for the lifetime of the platelet (Figure 4.3). TXA_2 is generated by the aspirin-sensitive cyclooxygenase pathway from arachidonate by way of the endoperoxides PGG_2 and PGH_2. Other NSAIDs are reversible inhibitors and baseline platelet response returns about 3 days after medication is stopped. The use of aspirin in conjunction with such other anti-platelet therapies as GPIIb–IIIa antagonists has demonstrated additive inhibition of platelet aggregation response to weak agonists such as ADP. Interestingly, there have been reports of individuals who are nonresponders to aspirin therapy. These patients not only have no inhibition of ADP and collagen-induced platelet aggregation but also have a higher risk of reocclusion following peripheral angioplasty.

Ticlopidine

Ticlopidine is a platelet function inhibitor that has been used increasingly in the area of cardiovascular disease. Ticlopidine mainly inhibits the aggregation response to ADP. Aggregation induced by such other agonists as thrombin or collagen is also inhibited to a lessor extent. The mechanism of action is unclear, as a metabolite of the parent compound exerts its inhibitory effect. It has been reported that ticlopidine inhibits reduction of cyclic AMP by ADP activation and inhibits binding of fibrinogen to platelets. It generally takes about 4 to 7 days after drug administration to demonstrate peak effects. Once treatment is stopped, inhibition of platelet aggregation may continue for more than 72 hr. The combination of ticlopidine and GPIIb–IIIa antagonists has been used in healthy volunteers and was demonstrated to have enhanced inhibition of platelet aggregation but did not prolong bleeding time. One potential side effect of ticlopidine therapy is neutropenia. A second-generation drug, clopidogrel, may be as effective as ticlopidine without the apparent risk of neutropenia. A recent randomized blinded trial of clopidogrel versus aspirin in patients at risk for ischemic events showed a 8.7% relative risk reduction with clopidogrel compared to aspirin.

DRUGS THAT MAY AFFECT PLATELET FUNCTION

Many prescription and over-the-counter medications can affect platelet function. It is important that a complete medication history is taken from each blood donor in order to interpret the platelet function test results as accurately as possible.

Antibiotics

Antibiotics that have a β-lactam ring structure, such as the penicillins and cephalosporins, may affect platelet function. The mechanism of action is postulated to be a membrane change that blocks receptor–agonist interactions, or affects Ca^{2+} influx.

Dipyridamole

Dipyridamole is a pyrimidopyrimidine that enhances the enzyme adenylate cyclase and causes elevation of cAMP. Elevation of cyclic AMP lowers the ability of platelets to become activated by platelet agonists.

Fibrinolytics

Fibrinolytic agents cause decreased fibrinogen levels and increased fibrin degradation products (FDP). Increases in the concentration of fibrinopeptide A in the plasma suggests continuing thrombin activity in patients treated with thrombolytic agents. It has also been proposed that platelet activation occurs at the thrombus undergoing lysis by thrombin liberated from the dissolving clot or generated by activation of the coagulation system. Therefore, platelets from these patients may either have reduced or increased responsiveness to added agonists during platelet activation or aggregation testing. Both potent antithrombins and GPIIb–IIIa antagonists are being evaluated as conjunctive agents.

Dextran

Intravenous infusion of dextran that may occur after coronary stent placement can result in reduced platelet function and may cause increased bleeding times.

Anesthetics

Anesthetics have been demonstrated to have an effect on the aggregation response of platelets. Anesthetics such as lidocaine, dibucaine, cocaine, etc., have a direct effect on the platelet membrane. Cocaine added to platelets *in vitro* causes a reduction in fibrinogen binding to the activated GPIIb–IIIa receptor. The concentration at which these anesthetics mediate a platelet inhibitory effect is an order of magnitude greater than that considered to have potentially lethal effects *in vivo*. There have been reports that patients undergoing general anesthesia not only had prolonged bleeding time but also had reduced aggregation response *ex vivo* to weak agonists.

Thrombin Inhibitors

Thrombin is a key regulator in the pathophysiology of acute coronary syndromes. It mediates the conversion of fibrinogen to fibrin, activates factor XIII that aids in clot stabilization, and is a potent agonist of platelets. Heparin is the most used clinical antithrombin, but it has many limitations in regard to efficacy and safety. Thrombin inhibitors such as heparin may cause heparin-induced thrombocytopenia (HIT) or thrombocytopenia in conjunction with acute arterial thrombosis. Heparin-induced thrombocytopenia with or without arterial thrombosis will generally occur 5 or more days following exposure to heparin. HIT is caused by an immunoglobulin that activates platelets in the presence of drug. The recent generation of direct thrombin inhibitors that act independently of antithrombin III are currently being evaluated in clinical trials. Two direct bivalent inhibitors, hirulog and r-hirudin, as well as two catalytic site inhibitors, argatroban and inogatran, are being compared for antithrombin efficacy. These direct inhibitors can inhibit clot-bound thrombin as well as inhibiting thrombin-induced platelet activation without the potential of inducing thrombocytopenia. Because thrombin plays a normal role in hemostasis, a major complication of thrombin inhibitors in clinical use is bleeding complications. For this reason, coagulation factors have been a recent target for development of new anticoagulants since thrombin formation can be interrupted without inhibiting platelet activation. Due to its role in both the intrinsic and extrinsic coagulation pathways, potent selective inhibitors for activated factor Xa have been developed.

Many other drugs have been identified that alter normal platelet function. For a comprehensive review, please refer to recent texts that describe these drugs in more detail.

CLINICAL TRIALS AND PHARMACODYNAMIC MEASUREMENTS OF PLATELET INHIBITION

In evaluating the effect of drugs on platelet function in multicenter studies, the procedures for platelet function testing should be as consistent as possible. In our experience as a core laboratory for several phase II trials, we have found that the method of venipuncture, blood processing, testing, and calculation of data are all critical to the success of studies. The variables of aggregation testing outlined in Chapter 2 also apply to pharmacodynamic measurements for these multicenter studies.

A comprehensive standard operating procedure must be outlined for the laboratory aspect of each trial. Details on venipuncture method, blood handling and processing, and testing methods should be clearly outlined. It is also recommended that the pharmaceutical sponsor provide the blood-processing materials, including labeled tubes and disposable pipettes as well as the agonists that are to be used in the aggregation assay. The agonists should be prediluted, frozen into aliquots, and designed as a single-use reagent. Generally, these materials are prepared in kits by the Core Laboratory and sent to the individual sites. In addition, initiation visits are performed by the Core laboratory staff to document proper coordination of the study, technical expertise, and instrument performance. When patient testing has been performed by the site at the specified sampling timepoints, the aggregation tracings or test results are forwarded to the Core Laboratory for evaluation and final reporting. Testing is routinely monitored, and sites are contacted if test results suggest a technical or instrument difficulty.

REFERENCES

Antiplatelet Trialists' Collaboration (1994). Collaborative overview of randomized trials of antiplatelet therapy, I: prevention of death, myocardial infarction and stroke by prolonged antiplatelet therapy in various categories of patients. *Br. Med. J.* **308**, 81–106.

CAPRIE Steering Committee (1996). A randomized, blinded, trial of clopidogrel versus aspirin in patients at risk of ischaemic events (CAPRIE). *Lancet* **348**, 1329–1339.

FitzGerald, G. A. (1996). The human pharmacology of thrombin inhibition. *Coronary Artery Dis.* **7**, 911–918.

Grotemeyer, K.-H., Scharafinski, H. W., and Husstedt I.-W. (1993). Two-year follow-up of aspirin responder and aspirin non-responder: a pilot study including 180 post-stroke patients. *Thromb. Res.* **71**, 397–403.

Jennings, L. K., and White, M. M. (1998). Expression of LIBSs (Ligand Induced Binding Sites) on GPIIb–IIIa complexes and the effect of various inhibitors. *Am. Heart J.* **135**, 5179–5183.

Jennings, L. K., Tardiff, B., Kitt, M., and Gretler, D. (1998). Comparison of receptor occupancy and platelet aggregation response of eptifibatide administered intravenously in patients with unstable angina or non-Q wave myocardial infarction. *JACC* **31**, 93A.

Jennings, L. K., Wise, R., Ramsey, K., Jacoski, M., Tamby, J. F., May, S., White, M. M., Cholera, S., Mendoza, C., Todd, M., Rush, J., and Giugliano, R. (1998). Comparison of platelet aggregation response and receptor occupancy of RPR 109891 administered intravenously in patients with recent acute coronary syndromes. *JACC* **31**, 353A.

Mueller, M. R., Salat, A., Stang, P., Murabito, M., Pulakl, S., Boehm, D., Koppensteiner, R., Ergun, E., Mittlbosck, M., Schreiner, W., Lasert, U., and Wolner, E. (1997). Variable platelet response to low-dose ASA and the risk of limb deterioration in patients submitted to peripheral arterial angioplasty. *Thromb. Haemost.* **78**, 1003–1007.

Phillips, D. R., Teng, W.S, Arfsten, A., Nanizzi-Alaimo, L., White, M. M., Longhurst, C., Shattil, S. J., Randolph, A., Jakubowski, J. A., Jennings, L. K., and Scarborough, R. M. (1997). Effect of Ca^{++} on Integrilin$^{™}$ GPIIb–IIIa interactions: enhanced GPIIb–IIIa binding and inhibition of platelet aggregation by reductions in the concentration of ionized calcium in plasma anticoagulated with citrate. *Circulation* **96**, 1488–1494.

Stein, B., and Fuster, V. (1992). Clinical pharmacology of platelet inhibitors. In "Thrombosis in Cardiovascular Disorders," pp. 99–119. Saunders, Philadelphia.

Stengert, K. B., Sellick, C. L., and Lazerson, J. (1982). Halothane-induced platelet dysfunction. *Anesth. Analg.* **61**, 217–220.

Umemura, K., Kondo, K., Ikeda, Y., and Nakashima, M. (1997). Enhancement by ticlopidine of the inhibitory effect on *in vitro* platelet aggregation of the glycoprotein IIb/IIIa inhibitor tirofiban. *Thromb. Haemost.* **78**, 1381–1384.

Warkentin, T. E., Beng, H. C., and Greinacher, A. (1998). Heparin-induced thrombocytopenia: towards consensus. *Thromb. Haemost.* **79**, 1–7.

APPENDIX

93

RECIPES FOR CITRATE-BASED ANTICOAGULANTS

0.1 M Buffered Citrate

    1.765 g sodium citrate (dihydrate)

    0.840 g citric acid (monohydrate)

    Bring to 100 ml with distilled water

    Use 1 ml anticoagulant : 9 ml whole blood

0.129 M Buffered Citrate

    3.2 g sodium citrate (dihydrate)

    0.42 g citric acid (monohydrate)

    Bring to 100 ml with distilled water

    Use 1 ml anticoagulant : 9 ml whole blood

ACD (acid citrate dextrose)

    2.5 g sodium citrate (dihydrate)

    1.5 g citric acid (monohydrate)

    2.0 g dextrose

    Bring to 100 ml with distilled water

    Use 1.4 ml anticoagulant : 8.6 ml whole blood

ACD-A (acid citrate dextrose formula A)

    2.2 g sodium citrate (dihydrate)

    0.8 citric acid (monohydrate)

    2.5 g dextrose

    Bring to 100 ml with distilled water

    Use 1.5 ml anticoagulant : 10 ml whole blood

INHIBITORS USED IN PLATELET TECHNIQUES

Compound	Final concentration
Prostaglandin I_2	1.0 nM
Prostaglandin E_2	0.05 μM
Indomethacin	10 μM
Benzamidine	1 mM
Staurosporine	200 nM
Okadaic acid	2 μM
Apyrase	1–10 units/ml
Pyruvate kinase	14.3 units/ml
Cytochalasin D	10 μM
Anti-GPIIb–IIIa	10–20 μg/ml
ASA	1 mM
Aprotinin	10 μg/ml
Leupeptin	10 μg/ml
Phenylmethanesulfonyl fluoride	1 mM
Hirudin	1 U/ml

CYTOSKELETAL PREPARATION FROM PRP

1. Withdraw PRP from aggregometer cuvette and immediately add 0.5 ml iced cold 2% Triton extraction buffer (PRP variety).

2. Vortex for 10 sec.

3. Centrifuge at 10,000g for 5 min.

4. Discard supernatant and wash pellet ×2 in extraction buffer.

5. Add 0.5 ml sample buffer to final pellet.

6. Heat for 10 min at 100°C.

7. For total platelet protein comparison, prepare whole platelet lysates by adding two drops 5 mM EDTA to PRP, then
   centrifuge at 1600g for 10 min. Remove supernatant and
   wash pellet ×1 in buffer containing 140 mM NaCl, 20 mM
   HEPES, and 1 mM EDTA, pH 7.1. Re-pellet and resuspend
   in buffer containing 137 mM NaCl, 2.7 mM KCl, 1 mM
   MgCl$_2$, 3.3 mM NaH$_2$PO$_4$, and 20 mM HEPES, pH 7.4, to
   half the original volume. Add equal volume of sample buffer for SDS PAGE analysis.

Cytoskeleton extraction buffer	PRP
2% Triton X-100	2 ml/100 ml
100 mM Tris	0.21 g/100 ml
10 mM EGTA	0.38 g/100 ml
2 mM 2-mercaptoethanol	14.1 µl (14.2 M stock)/100 ml
2 × 10^{-6} M leupeptin	100 µl (25 mg in 2.5 ml PBS)
pH 7.4	

CYTOSKELETON PREPARATIONS FROM WASHED PLATELETS

1. Lyse platelets with 2% Triton extraction buffer (washed
   platelet variety).

2. Centrifuge at 10,000g for 5 min to obtain cytoskeletal core.

3. Withdraw supernatant and recentrifuge at 100,000g for 2.5
   hr at 4° to obtain the high-speed pellet.

4. Solubilize pellets and final supernatant with sample buffer.

Cytoskeleton extraction buffer	Washed platelets
2% Triton X-100	2 ml/100 ml
10 mM EGTA	0.38 g/100 ml
100 mM benzamidine	1.56 g/100 ml
2 mM phenylmethylsulfonyl flouride	1 ml (35.8 mg in 1 ml 95% ETOH)/100 ml
100 mM Tris pH 7.4	1.21 g/100 ml

PREPARING GEL-FILTERED PLATELETS

Column Preparation

1. Put stainless steel filter support (Millipore XX30025-10) at bottom of 60 ml monoject syringe.

2. Place rubber ring over support to hold in place.

Electrophoresis sample buffer for cytoskeltons	
2% SDS	1 g/50 ml
2% 2-mercaptoethanol	1 ml/50 ml
10% glycerol	5 ml/50 ml
0.01% bromophenol blue	0.5 ml (1% in water)/50 ml
1 mM EGTA	0.019 g/50 ml
4 mM EDTA	0.076 g/50 ml
5 mM Tris pH 6.8	0.3 g/50 ml

3. Pour Sepharose 2B into syringe being careful to avoid bubbles between the filter and the bottom of the syringe (syringe must be capped at bottom).

4. Pour approximately 25 ml of HEPES Modified Tyrodes Buffer (HMTB) into syringe.

5. Thoroughly mix Sepharose 2B and buffer.

6. Place a transfer pipette filled with additional Sepharose 2B against one side of the syringe.

7. Uncap the bottom of the syringe and allow HMTB to begin to flow out.

8. When buffer gets to about 5 ml, pipette Sepharose 2B down the side of the syringe.

9. Add Sepharose 2B until column is full (50–55 ml).

10. Allow the matrix to just barely dry before beginning to wash with HMTB.

11. Wash column with approximately 150 ml HMTB.

12. Immediately before gel-filtering, wash column with 4× volumes of HMTB *with albumin.**

HEPES Modified Tyrodes Buffer (HMTB)	
12 mM Na bicarbonate	1.008 g
138 mM NaCl	8.065 g
5.5 mM glucose	0.991 g
2.9 mM KCl	0.216 g
10 mM HEPES	2.383 g

Put into 900 ml distilled deionized water; pH to 7.4; bring to total volume of 1000 ml with distilled deionized water.

*To make buffer with albumin, add BSA to a 0.1% final concentration.

Platelet Preparation for Gel-Filtration

1. Draw whole blood into ACD (1.4 ml ACD + 8.6 ml blood).

2. Centrifuge 135g for 20 min to obtain PRP.

3. Remove PRP to separate tube.

4. Layer PRP (5–6 ml) on column and allow to run into Sepharose gel.

5. Add HMTB with albumin.

6. Collect platelets and count; dilute to 250,000/mm^3 with HMTB.

7. Add 10 units apyrase/ml of GFP and allow to incubate at 37°C for 30 min.

8. Add 20 mM CaCl$_2$ and 20 mM MgCl$_2$ in a volume sufficient to yield 1 mM final concentration of both.

9. Incubate at 37°C for 15 min.

10. Platelets are now ready for use in assays.

NOTE: If gel-filtered platelets are used in aggregation studies with weak agonists, fibrinogen must be added; 200–400 μg/ml may be necessary to support aggregation.

PREPARING WASHED PLATELETS

Anticoagulants and wash buffers should be at room temperature (RT); platelet activation inhibitors may be added to anticoagulant and/or buffers, as desired.

Washed platelets can be prepared from freshly drawn blood or freshly expired platelet concentrates.

From freshly drawn blood

1. Draw blood and place in 15 ml plastic tubes containing ACD (8.6 ml blood to 1.4 ml ACD).

2. Centrifuge at 135g, 20 min, RT.

3. Carefully remove PRP with dispo-pipette (plastic) down to ~0.5 cm above buffy coat (do not disturb buffy coat or red cells). Transfer PRP to fresh 15 ml tubes (no more than 10 ml PRP per tube).

4. Centrifuge PRP at 1000–1200g for 10 min to isolate platelets. Carefully remove the plasma taking care not to disturb the platelet pellet.

5. Wash platelets with CGS (buffer may change for certain procedures) ×2 by adding a small volume of buffer (0.5 ml) and resuspending the pellet. Slowly bring the volume to 5 ml with buffer; centrifuge 1000–1200g. Repeat ×2.

6. Finally wash ×1 with ETS (buffer may vary according to procedure).

Once suspended in the final wash buffer and before spinning, count platelets so that they may be suspended at the appropriate count desired in the final buffer (e.g., lysis buffer).

If washing multiple tubes, pool pellet in last wash buffer to get more accurate count.

From Freshly Expired Platelet Concentrates

1. Mix bag containing platelet concentrate by inverting.

2. Cut tubing and pour platelet concentrate into 50 ml plastic centrifuge tubes (40 ml maximum/tube).

3. Centrifuge at 135g for 15 min to remove red blood cells.

4. Carefully remove PRP (do not disturb red blood cells) and transfer to new 50 ml centrifuge tubes (40 ml max/tube).

5. Centrifuge platelets at 1000–1200g for 10 min.

6. Remove plasma carefully taking care not to disturb platelet pellet.

7. Wash pellet ×2 with 40 ml CGS (or desired buffer) and ×1 with ETS (or desired buffer) and count as before, when resuspended in the final wash buffer. Centrifuge and resuspend in the desired test buffer at the desired count.

Centrifuge times and speeds may be adjusted for varying volumes of platelets being washed.

CGS Buffer, pH 6.5	
10 mM sodium citrate	0.38g
30 mM dextrose	0.60g
120 mM NaCl	0.72 g
Bring to 100 ml with distilled water	
Store at 4°C	

ETS Buffer, pH 7.4	
10 mM Tris	0.61g
150 mM NaCl	4.38g
5 mM EDTA	0.17g
Bring to 500 ml with distilled water	
Store at 4°C	

SDS GEL ELECTROPHORESIS

Principle

Proteins are solubilized in SDS-containing sample buffer and applied to a polyacrylamide gel to separate the proteins based on their molecular mass. The extent of mobility into the gel corresponds to the molecular mass of the protein. The separated proteins are detected by staining the gel with Coomassie brilliant blue. After destaining the gel, only the protein bands retain the blue stain, and the mobility of each protein can be compared to protein standards that are electrophoresed on the same polyacrylamide gel. The apparent molecular weight of any protein can be estimated by comparing its mobility with that of the standard proteins. A representative SDS–polyacrylamide gel of platelet proteins solubilized in reducing sample buffer is shown in Figure A.1.

Depending upon the molecular weight range of the protein sample, select the percentage of acrylamide gel that is to be prepared. If the proteins to be electrophoresed have a wide

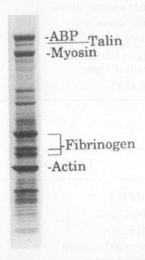

Figure A.1. Coomassie blue stained SDS–polyacrylamide gel of solubilized human platelets. Major platelet proteins are identified.

molecular weight range, a gradient gel (5–20% gradient of acrylamide) may be the best initial choice as it provides good separation of all proteins in the sample. If you wish to focus on a selected molecular weight range, then the lower the percentage of acrylamide in the gel, e.g., 5–7.5%, the better high-molecular-weight proteins (>70 kDa) will separate. However, if the proteins are in the middle- to low-molecular-weight range (25–70 kDa), a 10 or 12% acrylamide gel is recommended.

GEL RECIPES

Gel Reagents

Acrylamide/BIS

> 30 g acrylamide
>
> 0.8 g BIS
>
> In 100 ml Deionized Water (DW)
>
> NOTE: solution turns cold when dissolving
>
> Store at room temp

NOTE: Acrylamide is a neurotoxin: wear gloves and mask when weighing and avoid skin contact with solution.

1.5 M Tris-HCl, pH 8.8

> 18.16 g Tris
>
> Dissolve in 80 ml DW, pH, bring to 100 ml with DW
>
> Store at RT for 2–3 months

0.5 *M* Tris-HCl, pH 6.8

> 6.05 g Tris
>
> Dissolve in 80 ml DW, pH, bring to 100 ml with DW
>
> Store at RT for 2–3 months

10% SDS

> 10 g SDS
>
> Weigh carefully wearing particle mask as SDS is easily airborne and may irritate nose and eyes
>
> Bring to 100 ml with DW
>
> Store at RT for 2–3 months

20% SDS

> 20 g SDS
>
> Bring to 100 ml DW
>
> Store at RT 2–3 months

Separating Gel Solutions

Separating gel, 5%

> 28.45 ml DW
>
> 12.5 ml 1.5 *M* Tris-HCl, pH 8.8
>
> 0.5 ml 10% SDS
>
> 8.8 ml acrylamide-bis
>
> Mix well; just before pouring gel, add 25 µl TEMED and 250 µl 10% ammonium persulfate (make fresh each time)

Separating gel, 7.5%

> 24.24 ml DW
>
> 12.5 ml 1.5 M Tris-HCl, pH 8.8
>
> 0.5 ml 10% SDS
>
> 12.5 ml acrylamide-bis
>
> Mix well; just before pouring gel, add 25 μl
>   TEMED and 250 μl 10% ammonium persulfate
>   (make fresh each time)

Separating gel, 10%

> 21.75 ml DW
>
> 12.5 ml 1.5 M Tris-HCl, pH 8.8
>
> 0.5 ml 10% SDS
>
> 15 ml acrylamide-bis
>
> Mix well; just before pouring gel, add 25 μl
>   TEMED and 250 μl 10% ammonium persulfate
>   (make fresh each time)

Separating gel, 12%

> 16.75 ml DW
>
> 12.5 ml 1.5 M Tris-HCl, pH 8.8
>
> 0.5 ml 10% SDS
>
> 20 ml acrylamide-bis
>
> Mix well; just before pouring gel, add 25 μl
>   TEMED and 250 μl 10% ammonium persulfate
>   (make fresh each time)

Separating gel, 15%

    10.25 ml DW

    12.5 ml 1.5 M Tris-HCl, pH 8.8

    0.5 ml 10% SDS

    26.5 ml acrylamide-bis

    Mix well; just before pouring gel, add 25 μl
      TEMED and 250 μl 10% ammonium persulfate
      (made fresh each time)

Separating gel, 20%

    3.75 ml DW

    12.5 ml 1.5 M Tris-HCl, pH 8.8

    0.5 ml 10% SDS

    33.3 ml acrylamide-bis

    Mix well; just before pouring gel, add 25 μl
      TEMED and 250 μl 10% ammonium persulfate
      (make fresh each time)

Gradient Gel 5-20%

	5%	20%
DW	14.25 ml	1.88 ml
1.5 M Tris HCl, pH 8.8	6.5 ml	6.5 ml
10% SDS	0.25 ml	0.25 ml
Acrylamide-bis	4.4 ml	16.65 ml

Add 12.5 μl TEMED and 125 μl 10% ammonium persulfate to each and pour gel with gradient gel pourer.

Stacking Gel (for all gels)

    6.1 ml DW

    2.5 ml 0.5 M Tris-HCl, pH 6.8

    100 µl 10% SDS

    1.3 ml acrylamide-bis

    Mix well; just before pouring, add 10 µl TEMED
      and 50 µl of 10% ammonium persulfate

Tray buffer pH 8.3

    9.0 g Tris

    43.2 g glycine

    15.0 ml 10%SDS

    Dissolve in 1000 ml DW, pH, bring to 1500 ml
      with DW

Electrophoresis Sample Buffers

⅓× Non-Reduced (NR) Sample Buffer

    In 15 ml centrifuge tube dissolve
    1.0 ml 0.5 M Tris-HCl, pH 6.8
    1.6 ml glycerol
    0.32 g SDS
    0.32 ml 1% bromophenol blue
      (shake before pipetting)

    Bring final volume to 4.0 ml with DW and shake
    well to mix (may need to heat in water gently
    to dissolve SDS

    To use: divide sample volume by 3 and add that
    volume of ⅓× NRSB

    Example: to 1.0 ml sample add 0.33 ml ⅓× NRSB

⅓× Reduced (R) Sample Buffer

 10.0 ml 0.5 M Tris-HCl, pH 6.8
 16.0 ml glycerol
 3.2 g SDS
 3.2 ml bromophenol blue
 8.0 ml beta-mercaptoethanol
 Bring to 40 ml
 Aliquot (2 ml) in 15 ml centrifuge tubes and
  freeze at −20°C
 To use: divide sample volume by 3 and add that
  volume of ⅓× RSB
 Example: to 1.0 ml sample add 0.33 ml ⅓× RSB

1× NR Sample buffer (make fresh for each use)

 1.0 ml 0.5 M Tris-HCl, pH 6.8
 0.8 ml glycerol
 0.8 ml 20% SDS
 0.32 ml 0.05% bromophenol blue
 Bring to 4.0 ml with DW
 To use: Add equal volumes sample and 1× NRSB
 Example: to 1.0 ml sample add 1.0 ml 1× NRSB

1× R Sample buffer

 10.0 ml 0.5 M Tris-HCl, pH 6.8
 8.0 ml glycerol
 8.0 ml 20% SDS
 4.0 ml beta-mercaptoethanol
 3.2 ml 0.05% BPB
 Bring to 40.0 ml with DW. Aliquot (2 ml) in 15 ml
  centrifuge tubes and store at −20°C
 Add equal volumes sample and 1× RSB
 Example: to 1.0 ml sample add 1.0 ml 1× RSB

1% Bromophenol Blue

    1 g BPB/100 ml DW

    Shake before pipetting

    Store RT for several months

Coomassie Brilliant Blue Stain

    0.75 g brilliant blue R

    1363 ml DW

    1363 ml MeOH

    273 ml glacial acetic acid (add acid to water; prepare in hood)

    Mix and store in plastic container at RT

De-stain for Coomassie Blue

    300 ml MEOH

    450 ml glacial acetic acid5250 ml DW

    30 ml glycerol

    Mix and store in plastic container at RT

WESTERN BLOTS (OF ACRYLAMIDE GELS)

Principle

Proteins are electrophoretically eluted from acrylamide gels onto a matrix by electrotransfer. Time of transfer will depend on the thickness of the gel and the molecular weight of the proteins to be transferred.

Materials and Methods (wear gloves)

1. The gel to be blotted, as well as the blot buffers, should be at room temperature. Blot buffer is: 0.05 M Tris, 0.38 M glycine, 20% methanol in 6 liters deionized water and should be made

fresh for each use. (Add 0.1% SDS (wt/vol) for proteins >20,000 daltons.)

2. The procedure is outlined for an LKB blot apparatus. Pour approximately 1 liter of blot buffer in a large plastic dish pan. Place the back panel of the blot cassette in the bottom of the dish. Place one sponge on top of this, being careful to remove all air bubbles. Cut two filter papers the size of the sponge, dampen them with buffer, and layer on top of the sponge. Cut transfer matrix to the size of, or a little larger than, the gel or gel piece to be blotted. Wet the blot paper by first briefly laying it on top of deionized water in a tray and then submerging in the water for 2 min. Put the matrix paper on top of the filters in the blot buffer; then place gel on top of this, being careful to smooth out air bubbles between each layer of the "sandwich." Place two more pieces of filter paper on top of the gel. Complete the sandwich with the other cassette by hooking the prongs together appropriately. Rubber bands may be added to tighten the sandwich further. (There should be good close contact of gel sandwich pieces. Add more filter paper, if necessary, to achieve this.)

3. Place the heat exchanger (for cooling buffer) into the blot tank. Place the prepared cassette into the tank with the back cassette at the back of the tank (transfer membrane should be closest to the anode (red)).

4. Pour buffer into the tank to completely cover the cassette; place lid on the tank so that the anode is closest to the membrane.

5. Connect water supply to heat exchange and transfer at 0.7 amp for desired time (usually 1½ hr for higher-molecular-weight proteins and ~30 min for lower-molecular-weight proteins).

6. Remove cassette from tank and expose gel and blot. Remove gel, trim blot to exact size of gel with a scalpel, and stain with Coomassie blue to assess efficiency of transfer. Place blot in Immune Stain Buffer (3% BSA in 0.01 M Tris, pH 7.4) to cover. If immunoblotting, block with this for at least 2 hr.

7. Put blot in Immune Incubation Buffer (1% BSA in 0.01 M Tris, pH 7.4) to cover and add test mAb at recommended dilution. Incubate for 2–4 hr. Wash blot ×4 for 15 min each with Immune Incubation Buffer. Add second species-specific Ab (goat anti-mouse or rabbit, depending on first Ab) conjugated with horseradish peroxidase or alkaline phosphatase, at 1:2,000–1:20,000. Incubate for at least 1 hr. Wash ×4 with Immune Incubation Buffer and develop according to manufacturer's instructions.

Western Blotting Using Chemiluminescence Substrates

1. Run gel and blot in usual manner (do not use azide in any of the solutions for this technique; azide causes inhibition of HRP).

2. Block in Immune Stain Buffer containing 0.1% Tween for either 1 hr at room temperature or overnight at 4°C.

3. Hybridize with 1:30:000 dilution of 1 mg/ml test mAb (this concentration may vary with different mAbs) either 1 hr at room temperature or overnight at 4°C.

4. Wash ×6 for 5 min with PBS containing 0.05% Tween 20.

5. Hybridize with HRP conjugate 1:30,000 in Immune Stain Buffer either 1 hr at room temperature or overnight at 4°C.

6. Wash ×6 for 5 min with PBS 0.05% Tween 20.

7. Make up chemiluminescent substrate according to manufacturer's protocol.

8. Incubate blot for 10 min in substrate; then remove and wrap in clear plastic wrap or seal in bag.

9. Expose to film for 1 min for preliminary evaluation of signal. Final exposure times will vary depending on signal intensity.

10. If blots are stored wet, they can be stripped and reprobed.

Immune Stain Buffer pH 7.4

    1.21 g Tris
    9.0 g NaCl
    30.0 g BSA
    0.2 g Na azide
    Dissolve in 800 ml DW, pH, bring to 1000 ml
      with DW
    Store at 4°C for 1 week

Immune Incubation Buffer pH 7.4

    1.21 g Tris

    9.0 g NaCl

    10.0 g BSA

    0.2 g Na azide

    Dissolve in 800 ml DW, pH, bring to 1000 ml
      with DW

    Store at 4°C for 1 week

STAINING OF PLATELETS FOR FLOW CYTOMETRIC ANALYSIS

Whole Blood

Blood is added to an isotonic buffer at a dilution of 1:10 within minutes of collection. The dilution buffer may already contain the test antibody and a second antibody that will conclusively distinguish the platelet population from the other blood components during analysis. One must establish the correct dilution for optimal binding of each antibody. Generally, the tag and the test antibody will have different conjugated fluorochromes, for example, FITC and PE, respectively. Depending on the antibodies, the mixture is incubated for 15 to 30 min and then either diluted into PBS or 0.2% formaldehyde in saline at a dilution of 1:10. Platelets are identified by the light scatter properties and by the binding of the platelet-specific tag. The fluorescence intensity of the gated platelets by the test antibody is then measured. Data are reported either as percentage of positive platelets or as the mean fluorescence intensity.

PRP

After PRP isolation, PRP is added to tubes and antibody is added to the PRP. The mixture is incubated without stirring and then diluted with buffer and analyzed on the flow cytometer without any washing steps. When activation markers are analyzed, it is recommended that the PRP be diluted in PBS or appropriate buffer prior to addition of agonist or antibody to minimize the formation of platelet aggregates. After incubations, the sample is diluted for flow cytometric analysis as recommended.

1. Prepare PRP as usual and dilute to 2.5×10^8/ml

2. Add 100 μl PRP to as many microfuge tubes as needed for control tube plus antibodies being tested

3. Add 1 μl antibody (1 mg/ml) to respective tubes; mouse IgG or rabbit IgG as a negative control depending on test antibody used.

4. Incubate at 37°C for 30 min.

5. Remove 10 μl of each incubation sample and add to the tube containing 500 μl PBS and 5 μl of PE or FITC conjugated anti-mouse or anti-rabbit IgG (1 mg/ml)

6. Incubate for 10 min at 37°C

7. Perform flow cytometric analysis for Mean Fluorescence Intensity.

NOTE: This is a general method we use for
staining. Some antibodies require higher
concentrations in the incubation tube
to achieve adequate labeling.

SOURCES OF AGGREGOMETERS AND AGGREGATION REAGENTS

Bio/Data Corporation
155 Centennial Plaza
P.O. Box 347
Horsham, PA 19044-0347
(800) 257-3282

Chronolog Corporation
2 West Park Road
Havertown, PA 19083-4691
(800) CHRONOLOG

Helena Laboratories
1530 Lindbergh Drive
P.O. Box 752
Beaumont, TX 77704
(800) 231-5663

Payton Scientific, Inc.
244 Delaware Avenue
Buffalo, NY 14202
(716) 852-6213

Sienco Inc.
9188 South Turkey Creek Road
Morrison, CO 80465
(303) 697-4539

Index

Actin, 13
Actin binding protein, 13
Adenosine diphosphate (ADP), 17, 18, 55
Aggregometer stir speed, 35
Alpha granules, 2, 3, 15, 54
Anesthetics, 88
Antibiotics, 87
Anticoagulants, 28–29, 30–33, 94
    ACD, 31–32
    ACD-A, 31–32
    Citrate, 28, 29, 32, 94
    EDTA, 30
    Heparin, 30
    PPACK, 30, 33
Arachidonic acid, 56
Aspirin (ASA), 84

Bernard–Soulier syndrome (BSS), 72–73
Bleeding time prolongation, 73
Blood collection, 39

CD32, 8
CD36, 7–8
CD9, 8, 9
cAMP, 86–87
Chemiluminescence, 111
Clinical trials, 89
Clopidogrel, 86
Clot retraction, 7, 57–60
    PRP, 59, 60
    Whole blood, 58, 59
Coagulation, 20

Collagen, 18, 19, 55–56
Cyclooxgenase, 85
Cytoskeletal preparations, 95–97

Dense granules, 2, 3, 16, 17
Dextran, 87
Dipyridamole, 87

Epinephrine, 17, 18, 55

FcγRII receptors, 8, 20
Fibrinogen, 3, 36
Fibrinolytics, 87
Flow cytometry, 61, 62, 112

Gel electrophoresis, 102–109
Gel-filtered platelets, 97–99
Glanzmann's thrombasthenia, 7, 72, 77
GPIIb–IIIa, 3, 6–12, 57, 80–81
    Activation, 3, 10, 83
    Antagonists, 80–81
    Antibodies, 8, 9, 57
        D3, 9, 57
        mAb7, 8, 57
    Characteristics, 7, 9
    Conformational states, 10, 12
    Platelet aggregation, 6
    Signaling, 9–11
GPIb–IX–V, 3, 6, 9
GPIV, 7, 9, 19
GPVI, 9, 19

Hematocrit correction, 40
Hemolysis, 37

115